SHARI,

Lots of luck

Lavell

PAINT FINISHES

CHARLES HEMMING

Demonstrations by Peter Farlow

CHARTWELL
BOOKS, INC.

A QUILL BOOK

© Quill Publishing Limited 1985

First published in the United States in 1985
by Chartwell Books Inc., a division of
Book Sales Inc., 110 Enterprise Avenue,
Secaucus, New Jersey 07094.

ISBN 0-89009-909-X

This book was designed and produced by
Quill Publishing Limited
4-6 Blundell Street
London N7 9BH

Art director Nigel Osborne
Senior editor Patricia Webster
Assistant editor Jane Laing
Designer Ian Hunt
Illustrator Richard Phipps
Photography John Heseltine Ian Howes Jon Wyand
Index Richard Bird

Quill would like to thank Osborne & Little Limited for providing
wallpaper designs

Filmset by QV Typesetting Limited, London
Origination by Hong Kong Graphic Arts Limited, Hong Kong
Printed and bound by Leefung Asco Printers Limited, Hong Kong

CONTENTS

INTRODUCTION 6

1 CHOOSING A COLOR SCHEME 8

2 TOOLS AND MATERIALS 18

TROUBLESHOOTER GUIDE 28

3 PREPARING THE SURFACE 32

4 BROKEN COLOR 50

5 FANTASY DECORATION 82

6 FINISHING TOUCHES 118

GLOSSARY 140

INDEX 142

ACKNOWLEDGMENTS 144

INTRODUCTION

Paint has greater versatility than any other surface finish. It not only offers infinite variety of color and texture, it also affords protection to the surface — and does both at less cost than any other decorative material. Its chromatic and textural variety also allows it to evoke other materials, be they wood, marble or other polished stones, tortoiseshell or fabric; paint can be transparent or opaque; it can even offer images of objects that are not there.

This book is about paint and the effects it offers which may be achieved by the amateur with imagination, care and enthusiasm. Most of the decorative techniques described are applicable to furniture as well as to walls, ceilings and floors. For instance, marbling may seem the preserve of walls, but can add panache to tables or kitchen units. Spattering — or its more specialized variant, porphyry — is as effective on bathroom fittings as on coffee tables. Picking-out can reveal the shape of moldings, cornices, or turned and scrolled chair legs, that have been rendered formless by bad painting.

Because of its great variety and versatility – and ease of application – paint needs to be used with discretion. Over-spicing the cake is as easy when creating a color recipe as when mixing a culinary one. Using too many different effects in a room – on the walls or furniture – can detract from the overall look and become confusing. It's easy to avoid this error. In observing a room and the objects in it, take into account three things: look first at the shape of the room and its proportions, which will influence what you should paint and how; the fall of light, which will affect the colors you choose; and the structural condition of the room and its contents, which will dictate the preparation of the surfaces to be painted. For example, in a room with variously shaped woodwork panelling, it's a mistake to marble the walls and give a mottled, broken color pattern like rag-rolling to the wood; the clash of effects would destroy form and create visual chaos. Colors are important, too; those which make a north-facing room warm may make a south-facing room quite overpowering. Finally, the finish must be suited to its context: applying marbling to furniture that has the lightness of bamboo looks like the old Music Hall joke about the featherweight dumb-bell — totally unsuitable and so unpleasing to the eye.

To turn paint's versatility to your best advantage, it's useful to know something about the structure of color, so that you can mix paint; materials, and the tasks they are suited to; and how to prepare the surfaces you wish to paint. The first three chapters will lead you through these planning stages. The following three describe all the various effects and finishes you can get, where to use them and how to achieve them. With a little care, you can create a whole new look, from spattering a foot-stool to wood graining the floor of the hall.

1

CHOOSING A COLOR SCHEME

When choosing paint, the first priority is color. Paint comes in every hue of the spectrum, giving a range of patterns and tones beyond imagination. To make the best use of this variety and to derive the greatest pleasure from paint in interior decoration, it is necessary to know the basics of how color works.

All rooms, like all people, look different. Even in a skyscraper with identically structured rooms, no two are the same; this is because each room is at a different height or has a different orientation, so the light in each is different. Light affects the way that colors strike the eye, so suiting your choice of colors to the prevailing light is vital. It is color that gives a room its identity. There is almost no such thing as an ugly room; virtually any room, well painted, can look good. A beautifully painted room will look beautiful even if there's nothing else in it. (Indeed, over-furnishing detracts from a room, by confusing the color and the fall of light, and in extreme cases wrecking the proportions and form of the space.) It is color that gives a room brightness, warmth or coolness, drama, elegance or playfulness; it is color as much as shape that gives a room "atmosphere", for it is the fall of light and the luxury of our color vision that give the world visible form and ambiance. The sheer range and panache of color can intimidate people — when they do not take it for granted — and yet the basic principles of color are very simple.

COLOR THEORY

Most of us learn in school about the structure of color, set out on a color wheel, and then swiftly forget about it; so a brief resumé will be useful. There are two main sets of color rules: those of light mixing, which must be applied when laying one color over another as in glazing; and those of pigment mixing, which control results when physically mixing different colored paints. Some terminology may be useful: the primary colors — red, blue and yellow — are often referred to as hues. When a pigment is lightened with white, it is strictly called a tint, and when darkened with black, a shade; so pink is a tint of red, and mustard is a shade of yellow. In practice, though, on manufacturers' color charts and in magazines and in other publications, the terms "tones" and "tonal values" have become widespread to describe the level of lightness or darkness of a color. These terms are useful because manufacturers produce scores of variations of, say, blue, to which they give registered tradenames; all of them are really a basic blue — say, ultramarine — with different amounts of white or black added, and

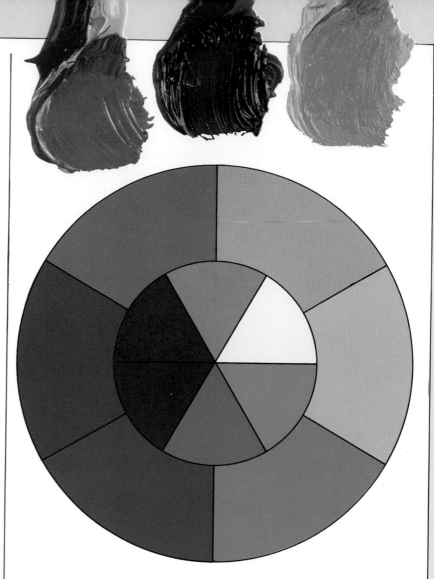

Above *The color wheel. In the center circle, the three primary colors — red, blue and yellow — appear with their secondaries between them. Orange is where red and yellow meet, green where yellow meets blue, and purple where blue meets red. In the outer circle, the complementary colors appear. Orangey-red, a color between tangerine and vermilion, is at the top left, and is a result of mixing the primary, red, with the secondary, orange. Magenta, on the left, is made by mixing red with the secondary, purple; and so on around the outer wheel.*

Mixing complementaries
Turquoise — mixed from the primary, blue, and the secondary, green — is at bottom right. If you mixed turquoise with tangerine, the complementary diagonally opposite it at top left, you would get gray. This would also happen if you mixed deep lilac at bottom left with ocher at top right, or magenta at far left with light olive at far right.

Top right *Large areas of primary colors used together dictate to everything around them but can be very striking.*

Right *The lighter effect of the blue dragged wood is strong without precluding the use of many other colors.*

sometimes both white and black added at the same time, giving a grayer blue. These are all tones. This term is useful, too, when you wish to describe a mixture of different colors in a room which all have the same intensity; for example, pink and sky-blue. They are different colors but neither is more intense than the other; they both have an equal amount of white pigment in them, so they are called tonally equal.

The primary colors — red, blue and yellow — are so called because you can't make them by mixing other colors. On the other hand, basic color theory tells us that from these primaries all the other colors can be made. This isn't strictly true, but all other colors contain two of the primaries and the primaries do produce the vast majority of colors. Red and yellow make orange; blue and yellow make green; blue and red, purple. These are secondary colors. If you mix two secondary colors you get a tertiary color. Olive is a tertiary, a mixture of purple and green. Tints and shades are produced by adding black and/or white to a primary, secondary or tertiary color. With the addition of black or white, of course, colors lack the brilliance or intensity of primaries and secondaries. Beige, for instance, is a tertiary paled with white.

Colors are usually arranged on a chromatic scale and displayed on color charts, but the famous "wheel" remains the most effective way of demonstrating the structure of color. Those colors which appear opposite each other on the wheel, or nearly so, are called "complementary". This is because when they are mixed in equal quantities they cancel each other out and, rather surprisingly, you get gray as a result. Mix red and turquoise in slightly different proportions and you will get variations of gray. The same thing will happen with blue and orange. (This is assuming that the pigments are pure — in fact, most commercial complementary paints are not pure and will produce a muddy khaki when mixed.)

When you place true complementaries side by side you get a "buzz"— the colors seem to overlap each other in a narrow, gray blur. This phenomenon can be useful in decorating; a harsh color of any type can be softened by adding to it a small amount of its complementary. A particularly stark red on a north-

Above *The different densities of pigment allow light to pass through paint and be reflected back, if the under-surface is pale. Here, a thin wash of red has been over-brushed with a denser layer in a cross-hatching pattern using a technique known as dragging.*

Right *Another example of using one color in different degrees of density. A shifting pattern has been achieved on this cupboard by brushing a wet gloss over a lighter tone of the same color and then using a crumpled rag to rumple the texture of the glaze so that the lighter tone shows through. This is known as ragging.*

facing wall can be softened to a warmer, weathered brick-red by adding a little turquoise, and orange does the same for a hard, chilly blue.

Of course, artificially produced colors available today have a brilliance which cannot be achieved by hand-mixing primaries and secondaries yourself. But if you seek to understand color, there is no substitute for mixing if yourself. You cannot, for instance, "think" a color. You can imagine scenes in color, of course — just as you may dream in color — but can you reproduce the color you imagine? Assuredly not. Colors are seen in relation to light and other contingent colors, so a color visualized in the abstract lacks a defining context. That is why when

you see a color in your mind's eye and go looking for it on a color chart, you so often can't find it; the color is only in the mind and is isolated from other colors around you. It has no physical existence.

Another excellent way of learning about color, especially if you are trying to decide on a color scheme for a room, is just to experiment with a water-color paint box, painting blocks of color and holding them up, sight-size (until they block out the area you want to paint) to the appropriate part of the room. This can save you time — and a fortune in paint. You can also discover interesting color combinations that you may have have been conditioned into thinking don't go together, like making delicate olives from purples and greens, or juxtaposing blue and green.

Of course, you don't have to mix the pigments of colors together to change them. It is generally admitted that color is at its most beautiful in its transparent state, applied over a white ground, with light shining through the color. Hold up a 35mm color slide to a light source and it is luminous, like stained glass, because the light shines through it. Almost all colors that artists use are transparent. Only a few are opaque — namely vermilion, cerulean blue, emerald green, ocher and some yellows. If you apply artists' paint directly from a tube and mix it with either water or mineral spirits, the original purity is never lost. Glaze painting is a technique involving overlaying one of these transparent colors with another or applying a transparent color over a white ground. It is a very ancient way of painting and retains the purity of the colors. Glaze painting is similar to the modern four-color printing process, where the various gradations of color are obtained by printing one color over another, on a ground of white paper.

Painting a picture in glazes is a more complicated process than painting by mixing pigments, but there are few such problems in using the technique for the paint finishes described in this book. All you need to know is that if you mix a transparent color and apply it over another color, that process is called glazing and there are certain formal color rules that it is useful to know.

In the glaze technique the aim is just the opposite from that in pigment mixing: transparency is essential for maximum effectiveness, so only transparent colors are used. Starting with a white ground, the painter covers the area with thin layers of transparent paint, which act as color filters. The white light shining down through them is reflected back up and out again from the white ground below. The more glazes that are put on, the less light is reflected back up, so the area appears progressively darker. You can go on until you get black but you will never get mud color; glazes retain a "clean" appearance down to the lowest levels of illumination.

In glazing, you are effectively removing various areas of white light; so you can no longer use the rules of pigment mixing described earlier, but must go by the rules of light mixing. To begin with, sunlight is white (strictly speaking a non-color) unless it is separated with a prism into the colors of the rainbow. When it is separated, it divides into pure colors; that is, colors that are undiluted and are therefore at their maximum intensity. These include the three primaries — red, blue and yellow. The primary colors for light mixing are orange (sometimes defined as red-yellow), green and blue. The complementary color of orange is blue-green, of blue is yellow, and of green is violet.

The basic rules of light mixing are: to soften a color, use a thin glaze of its complementary light color; to intensify it, add a second glaze of the same color. Blue over yellow still makes green, red over yellow, orange; *but* red over blue-green makes olive. So, for example, if you have an apple green wall you can give it a transparent glaze of rose color and create the same kind of mellow light that you see on a green tree on a sunny, late summer afternoon. If you mixed the same rose-colored paint in the can with green pigment, you'd get the kind of color you see on an army lorry. Similarly, glazing pure rose madder (deep rose) over a white ground gives a clear, deep rose. Mix rose madder paint with white pigment, though, and you'll end up with a paler pink. The result with mixed pigments is a surface color, not a transparent one; a color you can look *on*to but not *in*to.

USE OF COLOR

Because all rooms vary, there is no universally applicable recipe for a

successful interior; but there are more and less effective ways of using color. There is a simple way to observe a room and see what its problems are: walk about and look at all its different angles, paying special attention to the view from the door as you enter. Note the proportions: the height and width. Is the room too high or too low for its size? Are the doors symmetrically placed? Should they be made a feature, or should they be made to blend so as not to detract from the proportions of the room? Are the windows large, small, positioned oddly, too high or too low? Which way do they face? Look at the light from the windows: what will it be like at other times of the day? Will it be cold, clear, north light or warm southern or western? All these considerations should affect your choice of color and finish. If areas are broken with alcoves, hatches and recesses, or interspersed with windows of varying heights and widths, the room will benefit from unity of color and finish. The woodwork at least should be integrated with the walls, and not finished in a markedly contrasting color or style, otherwise the variousness becomes disturbing,

dictating to you rather than providing a pleasant environment.

When the quality of light is cold, such as in a dark, north-facing room, a common approach is to attempt to lighten and warm it with brilliant color. This frequently overwhelms the interior. What north light lacks is mellowness, so the harsh, deep reds — poppy and geranium — do not

their tendency to coldness can be offset with mustards, mellow terracottas and warm earth colors.

Sunny, southern-facing rooms can become too yellow; an excess of "sunny" colors can make them poky, like teapots full of tannin. They often benefit from cooler coloring; an almost-white pale pink or a very pale green can work very well.

This brings us to the one thing which all artists tend to presume is common knowledge — that some colors are considered cold and others warm . Blues and greens are usually the cool ones; reds and yellows warm. The reason for this is very simple. Color, as any astronomer will tell you, is divided into short and long wavelengths. Blue is a short-wavelength color, so it seems to recede; it looks far away , so a blue room looks cool because it reminds us of distance and space. There are visual associations too; the blue sky, the sea and distant mountain-tops are all blue, and cool. The more yellow blue has in it, the warmer and more summery it looks. Red is a long-wavelength color, so it seems to come toward us and looks close; it's therefore associated with warmth. Visual associations strengthen the sensation of heat — fire, the color of the red-brown earth, blood and life.

A lot of these associations are subconscious and very ancient, and it is easy to forget that because light is the cause of color, light changes color. To assume — as many people do — that because a color is associated with warmth it is automatically going to make a place look warm is logical but, unfortunately, not necessarily correct. Warmth can become heavy and coolness bleak. These effects are highly subjective.

There is no worse way of approaching a color scheme than as a rigorous intellectual exercise. The trouble is, we're conditioned by our society to take a logical, analytical approach to problems and to distrust our instincts and feelings as whimsical, self-indulgent, even as a sign of weakness. Choosing color, however, is all instinctive feeling in relation to light. "Over-think" is one of the reasons so many people have problems with color. They hear that dark blue is synonymous with dignity, strength and quiet dynamism , and so, psychologically, it is: it is a holy color to the Buddhists, not to mention the color of about half the world's dress

Above *The informal harmony of this room relies on the use of a single overall color. The walls are accented by very small, precise stencils.*

Far right *An example of tonal balance, based on ocher and a blue-gray of exactly the same visual weight, the gray being essential to balance the stark furniture.*

work well; they look bloody rather than warm in the gray light. Softer color, like brick-red, lightens and warms the room without being overpowering. A cold, green room can be glazed with red, or a red one with green, to produce amber-olives. Blues, greens and grays can all be darkened for depth and given a warm under-toning by choosing a shade with a touch of red or yellow;

uniforms, but that doesn't mean it's going to work well on a south-facing living room wall. Cover walls with it in that light, and you'll end up feeling like a parking ticket in a policeman's blue pocket. Similarly, if you paint a north-facing room scarlet, the cold light will make the color stark and it won't feel any more vital and warm than being walled up inside a fire engine; paint it bright lemon and you'll know what fruit juice in the freezer goes through.

There are two sorts of color relationship that it's wise to avoid, not because they are wrong but because they often cause other problems. This is what they are and how to remedy them.

OFF-HUES

When placed side by side, true contrasting colors do not alter your perception of their color; that is to say, vermilion red looks just as red, cerulean blue just as blue, and they also strengthen and intensify each other. On the other hand, colors more similar to one another but not actually adjacent on the color wheel — such as ultramarine blue and one of the red-purples, such as rose madder — send each other off-hue when they are placed side by side. These make very uncomfortable color combinations in a room. They can, however, be transformed by playing around with the proportions of pigment by which they're made and by using other tones that lie

Color Effects with Glazes

Most colors are transparent. The yellows, vermilion, emerald and cerulean blue are not. Glazing is painting in transparent color. If you thin paint with solvent — mineral spirits or water — this is called a paint glaze; if with glaze — a transparent gel available from paint suppliers — it is called tinted glaze. Paint glazes make semi-opaque washes; tinted glazes, translucent top coats.

between them; they can be linked by unusual and fascinating series of decorating colors. The difficulty with this is that if you have used large areas of contrasting colors, you will either have to use many smaller painted surfaces with the intermediary tones, or else use other materials to supply the link. Try to avoid large areas of true contrasting colors. They can be very disturbing in directly adjacent positions. On the other hand, they can be very effective if you have a large area of one, and the other in very small areas against it.

DISCORDANT COLORS

These are produced when the sequence of light and dark colors as they appear on the color wheel are unbalanced by adding an equal amount of white to the darker colors and black to the light ones, so that their relationships are reversed. A typical grotesque discord would be achieved by adding black to yellow — making mustard — and then adding white to a deep blue; used in equal quantities, this combination may turn your stomach and set you craving for tinted glasses. Again, the prime remedy for this is that very small areas of one color can work against large areas of the other — like a scattered pattern on fabric. Never use discordant colors in equal measure against one another.

Color is not a tyrant that cannot be placated until the personality of the individual decorating the room has been removed — quite the contrary The mere fact that color gives so vast a choice ensures that the choice reflects individual personality. So, too, does subtlety of color. Subtle colors are assertive precisely because they leave room for additions of stylistic panache and leave the options open for everything from quiet elegance to high drama.

Once you've settled on a color scheme, choose a finish that is suited to it, and to the surfaces you intend to paint. Wood graining a chair in navy blue and turquoise looks eccentric, to say the least; sponging a wall in olive and yellow will make you feel like you're under fire. As each technique is described, the pros and cons of its various uses are discussed. It's worth giving careful thought to the choice of finish and colors before you start, to ensure a pleasing, congruent unity in the finished effect.

Above Painted wood graining gives a crisp framing to this cool, varnished, off-white interior; the stark, clean lines of the furniture demand a visual balance on the walls and doors; plain painted surfaces would run the risk of becoming extremely heavy. Wood graining gives weight and variety without looking oppressive.

Right Ragging gives a damask-like or white velvet texture to these walls where a plain surface would be excessively neutral. But the walls must be "quiet", as the warm honey-ocher tones of other surfaces are gentle and discreet in themselves, while having a very distinct tactile quality.

Paint Mixing in Practice

Primary colors — red, blue and yellow — contain no white pigment. Colors which do are technically called tints or, if very pale, pastels . Colors made by adding black to a color pigment are shades , but today the term "tone" is often used to denote the brightness of a color. When mixing a pale color, it's best to start with white and add color to it. If mixing a dark color, start with the color (say red) and add black to it. Remember that it is easier to add color to a tint than to lighten a shade. If you make a color too dark, you won't make it lighter again by adding more color, just more intense. Toning down a color — *not* the same as darkening it — is most easily done by adding a small amount of its complementary; for example, adding turquoise to vermilion red.

2

TOOLS AND MATERIALS

There are few mediums more versatile than paint, and that versatility stems from paint's great variety. To take full advantage of this, you need the correct tools and an idea of how to use them. Different materials are suitable for different tasks, and this chapter surveys the various mediums and the tools associated with them.

Although there is a greater variety of paint available now than ever before, there are still only two basic types: water-based and oil-based. To put it simply, all paint consists of colored particles — pigment — held together in a binding solution, which is soluble in either oil or water. The great majority of water-based paints are not shiny when dry; about half of the oil-based paints are.

For the purposes of interior decoration, most people are primarily interested in what type of finish the paint will give, so manufacturers usually categorize interior paints according to the degree of shine they have when dry.

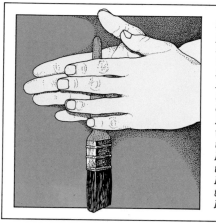

Left *Clean a brush as soon as you buy it. Twirl it with the handle between your palms, then give the brush a good wash in mineral spirits and twirl it dry. All brushes have loose hairs and twirling the brush, bristles downward, will cause the loose bristles to drop out now, not when you're painting. When using oil-based paint, clean brushes in mineral spirits and then wash them thoroughly in luke-warm soapy water; after twirling them, hang them bristles downward.*

Using Gloss over Large Areas

Woodwork and floors are the best surfaces for large areas of gloss, not walls. For shiny walls, use gloss varnish over flat-oil or latex. Finish is of paramount importance to gloss. When laying on, criss-cross the strokes to avoid tracks; then cross off feather-light finishing strokes with another brush, always going in the same direction. On wood, always follow the grain. On hardboard and similar surfaces, lay off toward the light. Don't use rollers; gloss tends to peel off when rolled, owing to insufficient adhesion.

Those with no shine are called flat or matt paints; those with a slight shine are low-luster, egg-shell, silk-finish or satin-finish; and those with a high shine, gloss or high-gloss.

Where to apply which paint is largely a matter of taste, but there are a few points to bear in mind. It isn't a good idea to put high-gloss or the more shiny low-luster paints on a wall with lumps or undulations because the paint will highlight the flaws. High-gloss paint on woodwork adjoining a matt wall draws too much attention to the woodwork and makes the walls appear to recede. Gloss paint washes more easily than other types, but woodwork tends to look better with low-luster or matt. If you need to protect paint from constant handling — on a banister or door, for example — it is preferable to varnish it. Three coats of varnish over matt paint are no more likely to chip than a thick coat of gloss paint; also, gloss has a hard, cold, colorless shine, whereas flat paint that has been varnished retains a gleam of its own color where it catches the light.

PAINT TYPES

MATT OR FLAT PAINTS

■ **Flat-oil** Widely used by professional decorators, this is the best paint for all interior surfaces; its consistency, coverage and finish are superior to all other paints. In the UK, flat-oil is available only through specialist suppliers. An undercoat is essential for a good finish with this and all other oil-based paints. Two thin top coats are better than one thick one; they are easier to brush on evenly and they adhere better. Using mineral

spirits, thin the paint until it holds to the brush like thin cream, but never more. Don't thin top coats as much. It is best to apply the paint fairly liberally, brush it to a thin, even film and then lay off carefully.

■ **Undercoat** Undercoat is an oil-based paint, used to provide a non-porous ground for all oil-based finishing coats. It is not usually applied as a top coat itself, but it can be substituted for flat-oil if necessary. If you do this, you may need two coats (as undercoat covers the surface more thinly than flat-oil) and you should protect it with a coat or two of matt varnish because its powdery texture makes it less durable than top coat paint. The color range is limited, but the colors are quite subtle and the paint is easily tinted.

Undercoat is an excellent base for over-decorating techniques, such as marbling, but you should buy one of the better quality brands for this, as they dry to a smoother finish. The first application of undercoat may be thinned half-and-half with mineral spirits to saturate a porous surface and to create a key for the next, full-strength coat.

■ **Latex** This term is used to describe a wide range of water-based paints which are usually applied to plaster; the range now includes matt vinyls also known as latex flat enamels. You cannot apply latexes to metal as, being porous, they will allow it to corrode. They aren't really suitable as a ground for over-decoration, either (unless sealed with a matt varnish), but they are good for washes because they can easily be thinned with water. Latexes are relatively inexpensive and are easy to use. They dry quickly and can easily be covered with other types of paint to avoid porosity. Lately, latexes have been produced to withstand steamy atmospheres, too.

You can apply latexes over new plaster, as they will let it breathe; they also have good adhesion and very little smell. A "mist" coat of latex on new plaster can be thinned 1:1 with water and the next coat applied at full strength. Apply this second coat liberally and don't over-brush. It's useful to lay off this paint toward a light source, rather than downward, to ensure an even appearance.

LOW-LUSTER TO GLOSS

■ **Egg-shell and low-luster** These can be oil-based or water-based. The oil-based paints are non-porous and give a soft, expensive-looking sheen on walls or woodwork. They are suitable for over-painting and other decorative techniques. They look much better when applied in several well-thinned coats rather than one thick one, which will look heavy and rubbery. The water-based paints are really designed for walls; they are fast-drying, but don't wear as well as oil-based ones. On woodwork, water-based egg-shells last longer than matt latexes. When thinned, their consistency should be no less than that of thin cream and they should be applied with the same brushwork method as flat-oil. Both low-luster and egg-shell paints need an undercoat.

■ **Trade egg-shell** Like flat-oil, this oil-based paint is usually supplied only to the decorating trade and is superior to all other low-luster paints. It is more expensive but more durable, hard and smooth with a better, even sheen and is suitable for doors, furniture and walls and as a ground for many over-decorating techniques. Although rather harder to work with than other low-luster paints, demanding careful brushwork, the effect of trade egg-shell is well worth the extra trouble. The methods of application are the same as for flat-oil.

■ **Gloss** All gloss paints are oil-based. The terms semi-gloss, gloss, high-gloss, wet-look and hard-gloss describe different levels of shine. Normally, the shinier the paint is, the more durable it will be. Gloss paints are highly resistant to water and dirt, although they do tend to chip. Gloss is usually applied to wood and metal and is often suitable for exterior use. It demands careful brushing on and laying off. These paints can be diluted with mineral spirits but they should not be thinned too much. The thinnest desirable texture is that of mayonnaise or tomato ketchup; any more thinning will break the consistency of the paint. Gloss paints need an undercoat and should then be laid on generously, over a small section of the surface at a time, and distributed evenly with cross-strokes. After this, tip off the brush on the edge of a paint kettle and then remove any excess paint. You can tell whether the application is even by the feel of the brush: it will slither greasily on saturated surfaces and drag over those that are too thinly covered. Use cross-strokes to correct any unevenness, and then lay off with firm, evenly pressed strokes. It's useful to apply successive coats in different directions to avoid a track

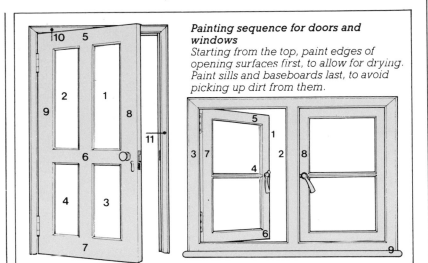

Painting sequence for doors and windows
Starting from the top, paint edges of opening surfaces first, to allow for drying. Paint sills and baseboards last, to avoid picking up dirt from them.

Using a paint kettle
The great advantage of using a paint kettle is that you use only a small amount of paint at a time, so that the remainder is protected from dirt and also from contact with the air, which will cause it to form a skin and become lumpy. Immerse only the first inch of the brush, and press the bristles against the sides of the kettle to remove brush on the rim, as paint will accumulate there and the lumps formed will drop into the paint. It's a good idea to wipe the rim and inside walls of the kettle at intervals with a dry brush to prevent any such deposits forming.

effect. If you are painting on wood, always finish the last coat in the same direction as the grain.

TINTING

Tinting means adding pigment to an already-mixed paint to alter the color, or adding color to varnish. The difference between ready-mixed decorating paint and pigments is that the first comes in bulk, ready for application, while pigments are colors sold in concentrated form. Pigments are soluble in either oil or water and can be thinned to different intensities. Never add pigment

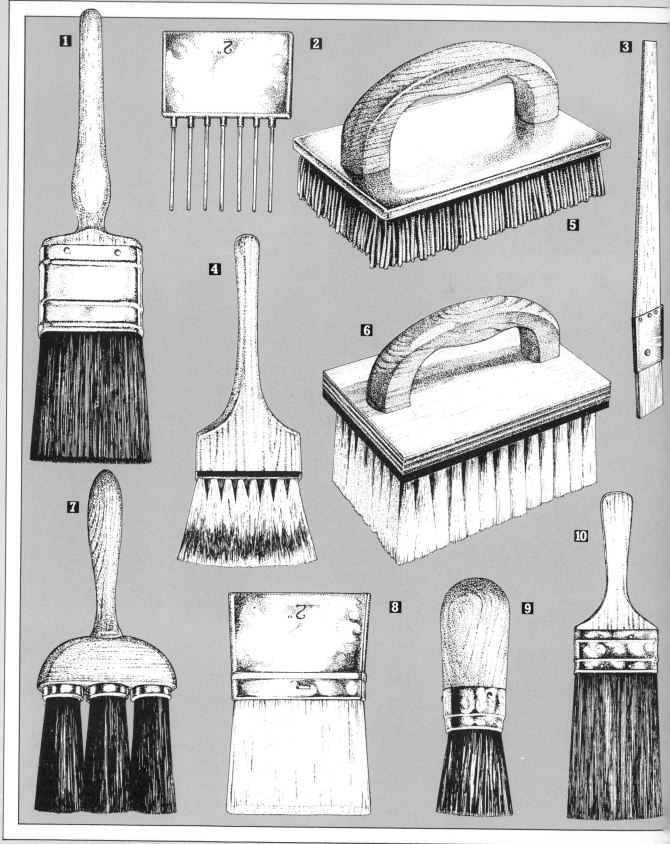

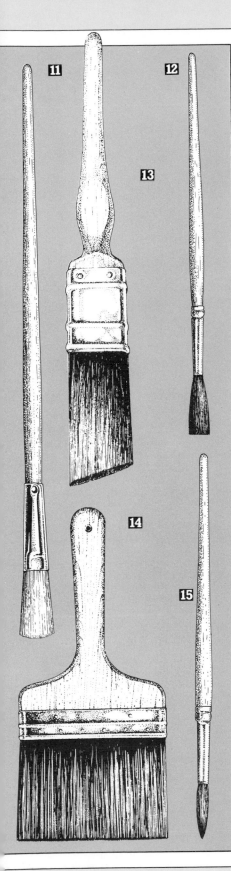

at a ratio of more than 1:8 to manufactured decorating paint, because the paint is already pigmented and may start to set. You should mix a sample after adding pigment and let it dry, because the color will alter as it dries. Mix the main quantity when you get the color you want on the dry sample.

OIL-BASED PIGMENTS

■ **Artists' oils** Oils can be added to any oil-based paint and are available from art shops. They are rather expensive but they offer by far the most sophisticated of all color ranges. The lowest prices are earth colors, such as brick-red and brown; the most expensive are the chrome yellows, cadmium reds and all the blues. Artists' oils are slow-drying and can be thinned with linseed oil, mineral spirits or turpentine. For tinting wall-paints, a small amount should be mixed to a creamy consistency with mineral spirits, using a palette knife, and then stirred into the decorating paint. Oils can also be used to tint varnish: when tinting cheap varnish, oils should be

diluted with a little linseed oil. Pigment spreads quickly in varnish, so only a little is needed. The pigment of cheaper oils is sometimes grittier than that of better quality ones. Good quality oils with a touch of linseed oil flow best.

WATER-BASED PIGMENTS

■ **Artists' powder pigments** These are soluble in water. They are available in fewer colors than the oils but they are strong and clear. They dissolve less easily in water than poster color (the other powder pigment), so there may be gritty particles that cause streaks; it's advisable to strain paint to which they've been added. They do, however, make excellent stains and washes.
■ **Poster colors** These are rather heavy, crude, powder colors with a large amount of filler but they are correspondingly inexpensive and easy to mix. The amount of filler means that they're not very concentrated and so the colors are less harsh when mixed into the paint.

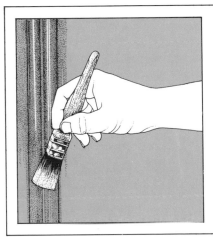

Laying-off strokes
Laying-off should be done with a nearly dry brush, using long, even, very light strokes. It is often advisable to lay off toward a light source, but on woodwork — whether on walls, floors or furniture — the direction of the grain takes precedence. The brush should be held like a pen, when painting wood; a flat brush is suitable for parallel moldings, an angled cutter-in for glazing bars or for fine details on furniture.

Left: Brushes
1 *Flat cutter*
2 *Seven-headed comb*
3 *Flat bristle liner*
4 *Badger softener*
5 *Rubber stippler*
6 *Bristle stippler*
7 *Three-headed softener*
8 *Dragger*
9 *Round stencil*
10 *Flogger*
11 *Bristle fitch*
12 *Sable writer*
13 *Angled cutter*
14 *Large standard flat*
15 *Artists' small sable*

■ **Artists' gouache** Professional artists, designers and illustrators use gouache. It is expensive but the colors are nearly as varied as those of artists' oils. Very concentrated and therefore opaque, gouaches make good stains and washes. In fact, they are probably the most versatile of all water-soluble pigments for washes.
■ **Artists' acrylics** These are very good, water-soluble, plastic-based paints and were originally developed as a counterpart to artists' oils, intended for the large, "sharp-edged" canvas paintings of the early 1960s. In practice, they offer a more

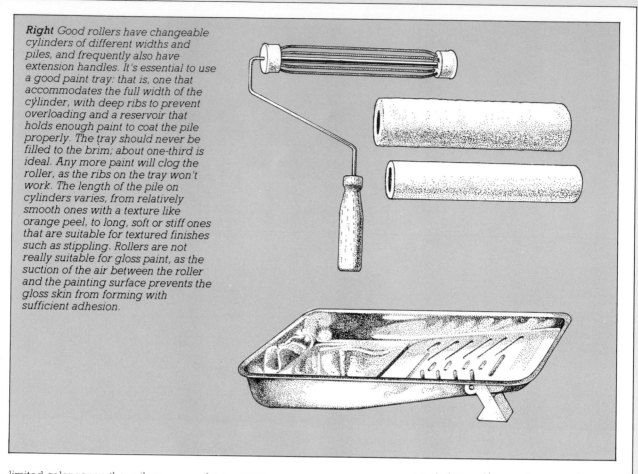

Right *Good rollers have changeable cylinders of different widths and piles, and frequently also have extension handles. It's essential to use a good paint tray: that is, one that accommodates the full width of the cylinder, with deep ribs to prevent overloading and a reservoir that holds enough paint to coat the pile properly. The tray should never be filled to the brim; about one-third is ideal. Any more paint will clog the roller, as the ribs on the tray won't work. The length of the pile on cylinders varies, from relatively smooth ones with a texture like orange peel, to long, soft or stiff ones that are suitable for textured finishes such as stippling. Rollers are not really suitable for gloss paint, as the suction of the air between the roller and the painting surface prevents the gloss skin from forming with sufficient adhesion.*

limited color range than oil or gouache. The range is limited in earth colors and neutrals, but is strong in mid-tones and pastels. Acrylics are not quite as expensive as oils, but you will need a greater quantity to paint the same area. They are very quick-drying; you can buy an effective retarder, but quick drying can be an asset in sharp-edged pattern-work like lining and stenciling.

■ **Universal stainers** These are used by the decorating trade to tint water-based paints. They vary from one manufacturer to another but are all very concentrated, so a little goes a very long way. The color range is limited and not very subtle but this is partly mitigated by the ease of mixing

All water-soluble pigments should be protected by a coat of clear varnish if they've been used as a wash over color already applied, so that they won't wash off. Some, such as the acrylics, are waterproof when dry and should not wash away, but it's always better to varnish a pigmented finish to be on the safe side.

EQUIPMENT

BRUSHES

Plasterwork is best painted with flat wall-brushes of between 4in and 5in (10cm and 12.5cm), which should be held midway along the handle, like a knife. For woodwork, including furniture, brushes are usually held by the stock, between thumb and fingers, and it's best to have 3in, 2in and 1in (7.5cm, 5cm and 2.5cm) cutting in brushes for window-frames, baseboards, details and moldings.

Always get the very best brushes you can afford, because the difference in performance between a good brush and a cheap one is enormous. It's also very important to look after them properly. Clean a brush as soon as you buy it, even if it's wrapped in cellophane; it's a dirty world and even the best brushes have loose hairs. Twirl the brush with the handle between your palms, give it a good wash in mineral spirits and then twirl it dry. Break new brushes in by using them for

priming and general preparation rather than for top coats, until all the short, loose hairs have come free. Brushes used with oil-based paint should be cleaned in mineral spirits, then washed in lukewarm, soapy water, rinsed and hung up to dry, bristles downward. Don't leave brushes with the bristles pointing upward because any remaining paint will drizzle deep into the hair roots, go hard, and cause a build-up of coagulated paint that separates the bristles until your brush looks like a molting cockatoo. Brushes used with water-based paint should be rinsed off in cold water first, then washed in warm, soapy water, rinsed again and hung up to dry. Never leave brushes resting on their tips or the bristles will distort. Store them flat when dry. If you want to leave them overnight to continue painting in the morning, oil-paint brushes can be suspended in a 1:1 mixture of mineral spirits and raw linseed oil, with the container covered to keep out dust and air; before use, rinse them in solvent and twirl them dry. For a similar limited period, latex brushes

Right *Broad- and narrow-bladed pliable spatulas are basic stripping tools. A triangular-headed shave-hook for corners and a curved one for rounded moldings are also useful.*

Below *Blow-lamps have a cooler flame than blow-torches but are less expensive over small areas.*

Spraying Furniture

Sprays offer a way of getting at recessed details that can be arduous with brushes, which may also lead to clotting paint. Unfortunately, sprays still need a primer and undercoat on stripped surfaces, although undercoat can be applied with a commercial spray. Aerosols are adequate for pre-painted surfaces and are quick, offering blending, mottling and frosty highlighting, effects that are difficult to achieve with brushes. Patterns can be made using masking tape. The visual height of a furnishing can be increased by darkening its lower parts and blending gradually lighter tones toward the top. This technique works well on a solid, four-square object; on a light, leggy furnishing it increases the air of lightness.

can be kept damp in a polythene bag.

When painting with brushes, always proceed methodically and don't rush; and use two brushes, so that one is clean for finishing. Go section by section (*see below*), laying on enough paint to cover a section, crossing the strokes to eliminate tracks and then laying off with a nearly dry brush in long, very light strokes in one direction (like using a feather duster) to get a smooth finish. Work sufficiently to ensure that the edges of the previous section are wet enough for you to brush the next one into it, but don't go back over a finished section or you'll ruin the texture.

Paint kettles enable you to use a small amount of paint at a time, while leaving the rest covered, which will protect it from gritty particles and prevent it from forming a skin. Don't immerse a brush up to the haft — it isn't necessary and clogs the bristles. Put the first inch or so of the bristles into the kettle and press each face of the brush against the side of the can to release excess paint. Don't wipe brushes on the rim, as this will form deposits that drop into the paint and make lumps. Use a nearly dry brush to take off any paint clinging to the inside walls.

ROLLERS

The advantage of rollers is their speed but this and the pile cause a superficial application of paint. This drawback is most pronounced with gloss paint, which can peel off in strips after roller application because of insufficient surface adhesion. For a textured surface, use a roller with a long pile; for general purposes, a medium pile; for very smooth surfaces, a short pile. As with brushes, buy the best you can afford. For high walls and ceilings, you'll need a roller with an extension handle.

The most common error in using rollers is overloading; an overloaded roller will spray paint everywhere like an exuberant wet puppy and squelch over the wall, leaving a texture like thin frogspawn. Avoid this by using a sloping paint tray and never filling more than one-third of its length with paint; any more, and the ribs on the bottom of the tray — which prevent overloading — won't work. After use, clean the roller in the appropriate paint solvent, then wash it in lukewarm, soapy water, rinse it and hang it up to dry. Like brushes, rollers used with water-based paints can be kept moist in polythene for short periods.

SPRAY GUNS

Large commercial spray guns are a swift way to paint and have long been used for decorating new houses in America. Spraying with latex over new plaster is particularly rapid, as the mist or sealing coat and top coats may be applied in quick succession. The primer and undercoat of oil-based paint take longer but the finish of oil-based paints is usually enhanced by careful spraying and application of the top coat is quick. As when using brushes, two or three thinned coats give a better finish and adhesion than one thick, so the fact that you have to thin paint for spraying is an asset. Spray guns vary in size and weight from small, bottle-fed, torch-like objects, reminiscent of science fiction ray-guns — which are light, convenient and highly maneuverable but give a small area of coverage — to big, canister-fed ones with nozzles in a flexible tube. Their use, loading and the desirable consistency of paint varies — follow the manufacturers' instructions. The main drawback of spraying is the time consumed in masking windows and fittings, although this is outweighed by the subsequent speed of execution. Also, a fine mist of paint may hang in the air, so it is advisable to wear a mask.

The best motion for spraying is a fluid rotation of the wrist or arm from side to side rather like a steady, featherweight brush-stroke. Keep the spray moving, like a blow lamp, only in wider sweeps; don't keep it pointed at one area or you'll get blotches and — unless you want a mottle — you'll have to respray the whole area. Also, such build-ups of paint can drizzle and stand out badly on the sprayed surface, and are difficult to eradicate if you allow them to coagulate and become viscous.

Sprays are so versatile for softer effects such as distressing and shading that it is tempting to use them beyond their capacity. Spraying leaves a surprisingly distinctive effect; it doesn't look good on marbling for instance, being too opaque and powdery. It can give a surface a cardboard appearance, lacking solidity, but for soft

Above left: Painting a ceiling
Work in strips parallel to the window, starting at the window side. Lay off with broad, crescent-shaped strokes, toward the light.

Above right: Painting walls
Work from the top in vertical strips, cross-brushing each strip as you go.

Unorthodox Equipment

Painting tools can be animal, vegetable or mineral. For mottled patterns, such as those found in marble, a sliced cauliflower floret has few equals. Chopped bottle corks make good wood-knot impressions. Cut carrots and potatoes are also useful for all mottled patterns, and crystals on porphyry can be done with diced cabbage, pressed flat by a tin tray.

To get whirling paint veins, quarter fill a rubber glove with paint, after piercing the fingertips, and squeeze from the top end. Pencils wrapped in cloth make convincing wood-grain strokes. The most versatile tool of all is your finger: twisted in a cloth to make knots, drawn across a wet glaze for a soft combing effect or in many other ways.

transitions and areas of smooth, unruffled color, spray guns have few equals.

GENERAL PAINTING SEQUENCE

The usual sequence for painting a room is: ceiling, walls, woodwork. When painting walls or ceilings, always paint the edges of sections first — where walls meet each other or a ceiling — using the tip of the brush, and try not to overlap the application into adjacent areas. If you are finishing furniture, it is advisable to do this after the structural elements of the room are complete. This way, you can be certain of matching, or complementing, the main colors of the room, which may alter as they dry.

■ **Ceilings** Start at the window side and work across in 2ft (60cm) strips, parallel to the window. Each strip is best painted in sections 2ft (60cm) square, first laying on strokes parallel to the window, then spreading the paint evenly with

broad, crescent-shaped crossing strokes and finally laying off toward the light. Try not to let the strips overlap when brushing on, but cross-brush and lay off to blend the two wet edges. Don't rush; go briskly but steadily.

■ **Walls** Start at the top-right-hand corner (or left-hand corner if you're left-handed) and work from top to bottom in 2ft (60cm) strips, parallel to the ceiling, cross-brushing each strip into the next as you go and laying off with a light, straight, downward stroke.

■ **Woodwork** Paint the window-frames, picture rail, doors, mantelpiece and baseboard in that order. This way, the areas that are lightest and cleanest on the brush are done first and the brush is therefore well broken-in but clean (with no loose hairs) before you get to the door — the most noticeable area for flaws, lumps and painterly hiccups. The baseboard comes last because the brush tends to pick up small, unwanted bodies near the floor.

On all woodwork, whether it's a door or a table, laying off strokes should follow the grain.

TROUBLESHOOTER GUIDE

PROBLEM	EXPLANATION	SOLUTION
How do I calculate how much paint I need to paint a room?	All good quality paints have their covering capacity stated — undiluted coverage equalling a given square metrage/footage — while paints which are sometimes referred to as "the produce of various countries" are very indefinite in density and often don't furnish such information accurately. Buy good quality paint. In the long run it is more economical and far more reliable.	Measure the base and one side of a square or rectangular wall and multiply to get the area in square meters/feet. Then consult the manufacturers' coverage details. Thinning will increase the area of coverage between 33% and 50%. Allow for one dilute and one full coat for both undercoat and top coats. Remember that you should not dilute gloss to a texture thinner than that of tomato catsup.
What floor coverings are necessary when burning off paint?	Burnt paint — especially oil-based — is very sticky. It rarely causes a naked flame but can easily scorch and leave gluey marks on wood, and bond itself formidably to carpet fibers. It is also very flaky, supplying its own tacky dandruff.	Aluminum foil is by far the most effective covering below an area of burning-off work. It is easily folded over, doesn't smolder, brushes off easily for re-use, and is light. Failing that, old chipboard (damp), bare floorboards if they're not going to show afterward, or wet sacking are suitable substitutes. *Not* paper. You should keep water and wet sacking in the vicinity as a matter-of-course precaution.
Having laid off away from the light, can I do anything?		Yes. *With latex* give another, thinned coat, laying off towards the light. *With silk or gloss*, allow the paint to dry thoroughly, sand it carefully and repaint it, laying off toward the light. Remember that, on wood, gloss should always follow the grain.
Can I remove brush hairs stuck to new paint?	On latex, loose hairs stick to the paint beneath them; on oil-based paint, under a film formed over them.	*On matt latex*, hairs can be sanded off, but a light patch may result. Alternatively, lift the hair with a razor blade, from the bottom upward, and lightly sand any little rim remaining. *On silk and gloss*, use a razor, sand lightly, and repaint if there is a noticable mark.
How do I know when gloss paint is really dry?		It takes some weeks. You can feel when it is fully dry: when it no longer feels like squeaky new polish but hard and smooth like plastic, it's dry.
Does emulsion have to be laid off toward the light?	Matt or egg-shell latex paint has a texture which refracts light, thus dispersing reflection.	Finishing strokes should go in a uniform direction, either straight down or toward the light. On a ceiling, lay off in bands parallel to the window, or you'll get a "floor-boards" effect.
Do I lay off with a roller in the same direction as with a brush?	The aim is to avoid patchiness and tracks.	The texture of roller-applied paint catches the light not like brush marks but like orange-peel. Criss-cross where necessary to get an even finish, but always lay off in one uniform direction. Don't go over parts that are drying, or you'll get an effect like old wallpaper.

PROBLEM	EXPLANATION	SOLUTION
I have a paint ripple on a flat gloss panel. Can I get rid of it?	Paint ripples are caused by excess thickness of paint. The top skin dries, and the paint often remains soft beneath.	On gloss — the type of paint with which this problem is most common — you can either take the paint off with solvent before it fully hardens and repaint, or let it harden, sand it off, and repaint.
How should gloss be laid off on a flat, grainless panel like chipboard?	What you want to avoid is a shiny, cross-hatching effect or a series of tracks, from thick paint catching the light.	Work quickly — but don't rush — in a close, criss-crossing stroke to avoid tracks, *then* cross off toward the light and smooth with a dry brush. Don't go back over drying areas.
Is it dangerous to breathe in sprayed paint particles?	The dispersion area from small aerosol cans is very small, and their limited capacity means that they are used close to the surface. Even so, used in a confined space in warm air, inhaled particles can give you a headache. Large sprays with a wide dispersion cast particles above and below you, and these should not be inhaled.	In any confined space, it's better to wear a mask. Small aerosols used briefly — to spray a chair leg, for example — don't really matter if there is proper ventilation. You should wear a mask for all other tasks. (Ecologists should remember that aerosols are reputed to damage the ozone layer.)
How does ventilation affect the effectiveness of sprays?	Any draft near a small spray will scatter it. Large sprays become less accurate and spraying takes proportionately longer.	Without a mask you need plenty of ventilation, and this scatters spray. Wear a mask, and ensure ventilation *without* a draft.
How much masking of surfaces is necessary with sprays?		Gravity causes sprayed paint particles to sink slowly. Large sprays necessitate covering floors and any furnishings and glass; small sprays, only the immediate 3ft (1m) or so. A ventilation source behind the spray is best. Paper or polythene sheeting is the most effective covering agent.
I have applied new gloss paint to wood and it has flaked off in places. What causes this and what can I do about it?	Peeling can be caused by: 1. Paint that is too thick and has dried too quickly. 2. Paint insufficiently keyed to the undersurface. 3. Heat. 4. Damp.	In all cases, scrape the peeling paint off carefully *(see chapter three)* from the affected area and sand down the edges of the remaining paint. Then: 1. If the paint is too thick, re-apply it more thinly, blending the edges into the surrounding paint with a dry brush. Re-apply the undercoat if this has been removed in stripping. 2. Insufficient keying is usually due to defective priming or defective undercoating. Patchy priming can cause damp, or resin bleeding through. Grease on the undercoat can cause peeling. Re-prime if necessary, wash and sand the undercoat, *let it dry*, then give another coat before re-applying the top coat. 3. If possible, remove any direct heat source. 4. Damp wood should never be painted. It should either be stripped and the damp treated, or stripped and repainted with latex, which lets it breathe.

PROBLEM	EXPLANATION	SOLUTION
Is it possible to remove sprayed paint from clothing?	Paint from small aerosols is usually enamel, and this leaves a film on fabric like nail polish. Oil-based and latex paints for walls and woodwork stain into the weave of natural fibers but will wash out with solvent and detergent if you act quickly.	If you do it promptly, *oil-based* paint can be sponged out with mineral spirits, then washed in water and detergent as hot as is appropriate to the fabric. *Latex* should be sponged off immediately with luke-warm water and then, if necessary, washed. Clothes which need dry-cleaning shouldn't be worn for painting work.
Can sprayed paint be removed from glass?		Slice sprayed spots off with a sharp spatula run flat over the glass. If the paint haze is very small and remains, rub vigorously with mineral spirits on a soft cloth or sponge, then scrape with the side of a palette knife. Finally, wash the glass with window detergent and polish it.
How can I paint a window frame without using masking tape, which leaves marks on the glass?		Use a metal rod with a raised rim and a handle. This is known as a ''George''. They are usually available from suppliers to the decorating trade.
Why does my brush look like a molting cockatoo and can I salvage it?	A mangy-looking brush is the result of: 1. Washing a brush in paint solvent and leaving it bristles upward, so that paint runs down into the haft and hardens in layers in the bristle roots, spreading them out. 2. Leaving a brush lying flat when not properly rinsed of solvent, so that one side dries hard and the other sticks up. 3. Leaving a wet brush resting on its bristles. 4. Not properly cleaning a new brush, so that paint coagulates along the hairs and glues them together.	First, soak the brush in paint remover and solvent, usually in a ratio of 1:1 parts remover to mineral spirits, for about two days. (This goes for a brush hardened with water- as well as oil-based paint.) When the bristles are softer, wash the brush thoroughly in warm, soapy water and then repeat the remover/solvent soak. Rinse it again after two days, and comb it with a stiff plastic or steel comb. If there is a paint skin like gritty scum on the bristles, lay the brush flat and run a razor blade in a flat, brisk, single-stroke, stropping motion, sweeping from the haft to bristle tips. Don't slice or saw or you'll chop the bristles off. Work with razor and comb, repeating the soaking and rinsing until the brush is soft and fluffy.
Radiators are frequently chipped and sometimes cracked. How can they be redecorated?	Because most radiators are enamel-finished they can chip if knocked. Peeling is caused by gloss paint that is too thick and does not conduct the heat, or a very hot radiator and/or one that's been painted without proper primer.	If chips and cracks are widespread, strip the paint with chemicals. If there is rust, clean it off *(see chapter three)*. When stripped, keep the metal *hand-hot* but no more, and prime it with zinc chromate. Then repaint it with oil-based non-gloss, as gloss's skin doesn't conduct heat well. *Never* repaint radiators with water-based paint.
What is the general effect of particular lighting positions?	Low moderate sidelights cause lengthening of shadows upward, and increase the sense of height and warmth. High moderate sidelights give a sensation of distance and depth downward but often give a rather bleak sensation. Strong central light in a small room is harsh and flattens everything. Softer central lighting needs soft side lighting, too.	

PROBLEM	EXPLANATION	SOLUTION
If a color has faded in patches on a radiator or on wood, can I touch it up so that the faded spots match the other color?		No. Paint which has begun to fade in patches will continue to do so. Other parts will fade later, and are already fading. What you matched will not match after a time.
Will some parts of a room fade more quickly than others?	Light is responsible for the fading of walls directly facing windows, and the lower walls and baseboards. Cigarette smoke rapidly causes discoloring of upper walls and ceilings. Painted furniture fades from wear more than from the light.	
Do some colors fade more than others?	Traditionally, yes. Manufacturers claim that today's commercial interior paints fade very little, which means that they are usually going to need repainting from general wear before they've faded — usually about four years. As many are very pale anyway, there is a loss of "sharpness" in them more than obvious fading.	Few commercial decorating paints remain unreplaced long enough for fading in the manner of pictures. So the resulting list works as an antiquing guide; over about ten years one might expect the following changes: Bright red fades to pink, then orange. Deep red fades to orange or brown. Dark blues fade to gray. Mid-blues fade to green. Light yellow fades to cream. Deep yellow fades to amber. Magnolia fades to brown/cream. Zinc whites fade to gray. Greens fade to gray.
Does varnish postpone the fading of color?	Yes, to a degree. The prime function of varnish is to protect the actual paint surface from damage, rather than to affect the chemical balance of pigments.	Polyurethane varnishes remain clear for many years, though they frequently lack a sense of visual depth. All varnishes darken in time. Paint that is mixed with varnish therefore fades very slowly, and is highly durable. For purposes of interior decoration, varnish/paint mixes might be regarded as the most stable-colored of finishes, provided that the varnish has been properly applied and isn't too thick. Broadly, varnish darkens and pigments fade, but unless you're intending the decor for your grandchildren you can expect no perceptible change.
How is changed lighting going to affect a room?	Generally speaking, artificial lights give yellow illumination and sunlight white, frequently with a blue tint. Moderate artificial light causes softening of color, exactly resembling the effect of an amber varnish. Sunlight heightens color.	Make color samples and view them in false and natural light to see how they change. Unless you know a room is to be used almost exclusively at a particular time of day and know its lighting accordingly, choose the tone which looks best in both natural and false light.
What causes fading of color?	Fading is due to light, dirt and heat. Light bleaches color. Dirt darkens and obscures it. As dirt can be washed off, older paint which is clean appears generally paler. Heat dries paint, causing a change in the pigment balance. This causes fading. Paint with a lot of oil in it darkens as the oil dims, but the pigment grows paler.	There is nothing you can do about paint fading with time, any more than you can prevent yourself growing older. In paint, fading is a sign of age. It can be beautiful.

3

PREPARING THE SURFACE

There are few surfaces in a domestic interior that you can't paint, but those that you can should be thoroughly prepared. This chapter shows how to remove old paint, paper or distemper quickly and easily, and explains how to prepare a wide variety of surfaces for painting, including plaster, wood, paper, fabrics, metal and tiles.

Removing distemper

1   First, brush thoroughly with a dry brush to take off any loose "dandruff" flakes from the surface.
2   Then thoroughly soak the surface with warm water, rubbing vigorously with the stiffest brush you can obtain. Change the water regularly, as soon as it becomes milky.
3   Once all the distemper has been removed, rinse the surface thoroughly with clean water and swab it all over with a sponge or soft cloth.

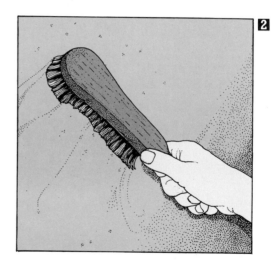

Paint is more versatile than any other surface finish and can be very durable; the animals sketched in earth and resin on the walls of caves have been with us for over 5,000 years. You can ask extraordinary things of paint; one of the few demands it makes in return is that you prepare a surface for it first. Although this may take time, you will be well rewarded if you do it well. So when you think of painting an interior, it's as well to know where it will go easily and where — in the minority of cases — it won't.
You can paint on good paint in good condition, any primed wall in good condition, new plaster and old plaster properly primed, sealed woodwork, sound varnish, lining paper or the backing paper of vinyls, primed fabric, painted or scoured metal, glass and tiles.
You should not paint on peeling paint, crumbling plaster, unsealed plaster, polyurethane varnish or old wallpaper.
You cannot paint on old distemper, unsealed wood, unprepared metal or felt.

   Such widely used metaphorical expressions as "painting over the cracks" and "doing a whitewash job" are a reminder that paint is only a surface finish. It can cover a multitude of sins but without proper care it can betray them or even make them appear worse. Few things are more depressing in an interior than seeing the joints of old, dried paper as it sags under the weight of new paint, high-gloss paint slapped over window sills with old paint blisters, an acne of trapped dust, unfilled (or, worse, unkilled) wormholes, or little moon-crater chips, all highlighted by the new, shiny surface. The preparation of a surface is, thus, very important.

OLD PAINT

The first thing to remember about paint in interior decoration is that it is not a substitute for structural integrity. If a wall is flaking or a piece of furniture is rotten, it's no good hoping that the paint will glue it together. On a sound surface, however, paint acts as a protective as well as an aesthetic agent. Unless the existing paint is visibly bubbling or peeling, or is damp — which will be betrayed by dark stains like sweat marks (because they actually are sweat marks) or, on light-colored paint, by yellowish stains like nicotine — it is usually a very good idea to leave the surface untouched, even if you wish to paint over it. Paint applied over sound paint usually retains as good a finish — if not better — as that laid over a stripped surface, and an extra coat only increases the protection.

■ **Distemper** The absolute exceptions to this are distemper, whiting and lime-wash. These can't easily be over-painted without "bleeding" or "dusting into" the new paint on top. They are usually found on older surfaces, particularly in wooden houses or the lower walls of tenement blocks. These surfaces do have to be removed, but fortunately that's quite easy. (After all, the reason you have to take them off is that they tend to come off.) A good way of testing for them if you aren't sure what they look like is to dampen your finger and rub it along the wall: if the paint marks your finger, it is one of the distempers. Distemper may come away similarly on dry fingers, like thin chalk; if that happens, it really has to be taken off.

To remove distemper, whiting or lime-wash, brush vigorously with a dry brush to take off all the dust and "dandruff" loosely adhering to the surface and then soak the surface thoroughly all over with plenty of warm water, scrubbing with the stiffest scrubbing brush you can obtain. The water will go cloudy very quickly, like thin milk; change it as soon as it does. Rinse the area thoroughly with clean water, and swab it all over with a sponge or soft cloth. You should leave the wall to dry out properly before you put on any type of paint, primer or sealer, or you will get damp underneath the paint.

■ **Other paints** All other painted surfaces, if they are in good condition, should just be washed down with ordinary soapy water, regardless of what type of paint you intend to put on top. Make sure you remove any grease or smudging, crayon or ink from them; ink washes off and things like wax come off with a putty rubber (available from any art shop). Once you've washed the surface with a sponge or cloth, let it dry. Then you can paint on it. When you paint over previous paint, you shouldn't need a primer or sealer because the surface is already sealed. You will in many cases need an undercoat, though, because its prime function is to block off the color that you are covering up so that it won't show through the top coat. Undercoats are described in detail in chapter two.

In the instructions on the back of a can of paint, practically all manufacturers state that loose and flaking paint should be removed before new application. If you do have to remove paint, either from wood or plaster, there are three courses of action open to you. Oil-based paints and latexes are both rather difficult to remove, which is a good reason to leave them there unless they're bubbled, wrinkled or cracked. If the paint is dishevelled in only one or two places, it's better to take it off just from these patches and sand down the edges of the stripped area until they are flush. On wood or plaster, it's best to see if the paint comes off by scraping before you try any other method. Frequently it will, especially if there is a build-up of thick coats and shrinkage has separated them from the surface beneath. Put a knife, a stripping knife or a spatula down behind them, if there's a gap, and run it along the surface, keeping it as flat as possible. If there is no gap, push the knife or spatula into the brittle part of the paint as if you were scraping grease off a plate, again keeping it flat so as not to score the wood or plaster beneath. If the paint comes

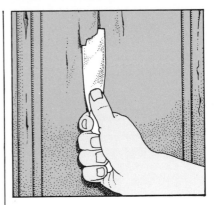

Above *Cracking and blistered paint should be tested by running a spatula or stripping knife along the surface behind the paint, keeping the blade as flat as possible. Sometimes no more than this is necessary to remove very thick, dry paint that has shrunk.*

Removing Paint and Varnish from Intricate Woodwork

Stripping turnings and recesses of wooden furniture is best done with chemicals, as a naked flame can char delicate details if held too close while a scraper picks at them, but the little electric strippers are quite useful for this sort of job. With chemicals, *allow them time to work*. Then, assemble a convenient combination of tools: a small shave-hook (used first to dislodge any lumps); a skewer; a wire-brush, a toothbrush or steel wool. Work down from the top, to avoid sticky bits of paint dropping into crannies already cleaned. You may need three coats of stripper but that's better than scraping too hard at unsoftened paint and damaging the detail. Wash the surface thoroughly when completed, either with a hose and then sponge, or soft brush and sponge. Most varnish removers should be used in the same way but they vary, so read the manufacturers' instructions.

off, work like this over as much of the area as you can. If other parts obviously ought to go but won't come off this way, then you will have to strip them either by burning or with chemicals. Both these methods can be used on furniture as well as on walls, floors and so forth. Methods of application are the same.

CHEMICALS

Latex paints should generally be removed with chemicals. Chemical paint- and varnish-removers have the advantage of not damaging the surface of plaster or wood, although they can damage clothing and rot shoe-stitching: walk carelessly in their splashes and the soles of your shoes may fall off. They should therefore be used with circumspection for any item involving fabric, such as a chair with an upholstered seat. They are definitely messier than burning-off and they can be more expensive over a large area, too, as you may need more than one application over layered paint. They divide into two types: spirit and alkaline.

Alkaline strippers are very powerful potions and highly effective. They work much more quickly than the spirit type, and are used by industry, professional decorators and artists — sculptors and painters alike — who use them to get special effects. Their strength means that they can be dangerous on the skin and both you and areas not to be stripped must be covered to prevent damage from splashes. They are also more difficult to wash off than spirit-based strippers and, because they are absorbed by wood, they can alter its grain, giving it a raised look. They also rot the hairs of ordinary bristle brushes to a ragged stump. In short, alkaline strippers are very good if you have plenty of experience of them and their properties but for an ordinary domestic interior it's easier and safer to use spirit strippers.

Spirit strippers are normally intended for domestic use, and some are even suitable for use on plastic. Like the more powerful alkalines, they are chemical solvents that soften the paint so that it can be scraped off. Like the alkalines, they are highly toxic, so be sure to work in a well ventilated place. Until recently they were all inflammable, but certain non-flammable brands are now available, which also means it's easier to get rid of the containers.

■ **Tools** A shave-hook, which looks like a trowel at right angles, and a broad-bladed, flat knife are essential for stripping. The best shave-hooks are lozenge-shaped with one straight edge, one curved and one deeply hooked for crannies. Steel wool, toothpicks or sausage sticks and a couple of old (preferably stiff) toothbrushes, one with a broken-off handle and one with a long one, are very useful for details, plaster moldings or wood. It's advisable to wear rubber gloves, too. The paint stripper should normally be applied with an old paintbrush; even the mildest chemicals will slowly eat brushes. Rags and newspapers are useful, especially for covering floors, and a metal container for collecting paint flakes and scraping tools on is very handy, too.

■ **Application** If you are the type of person who likes to finish a race almost before the starting gun has sounded, chemical strippers are not for you. You do need a considerable degree of patience, not because they're ineffective but because if you wait for them to take effect properly, the actual removal of the paint will be quicker and simpler. In almost all instances, you should brush the stripper on and leave it for at least half an hour. If it is a paste-textured stripper, don't brush it back and forth; put it on in one direction only, because otherwise you'll disturb the film and it won't work properly. Liquid strippers benefit from frequent applications, always keeping the surface wet. In either case, put the first coat on, let it soak in until the paint surface begins to soften, then put a thick second coat on. Then leave it alone. It may take hours to get through a thick build-up of paint — it may even take a couple of days — but that doesn't matter, since, when it has softened right through, the paint can just be stripped straight off like soft pastry, with no hard rubbing, leaving the surface beneath quite bare, clean and undamaged. It really is an awful waste of time and money to rush the job; you'll end up with a tacky, sticky molasses that needs hours and hours of scraping. When you are removing the paint, keep the edge of the tools as clean as possible by scraping them on a can rim or other edge and rubbing them on rags soaked in detergent. Wash the rags and steel wool in detergent, too. When the paint is entirely removed, just wash down the surface with either water or mineral spirits, whichever is the

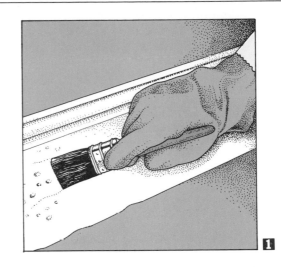

Application of chemical strippers

1   *Lay liquid stripper on in layers with an old brush, keeping the surface wet until the paint begins to bubble and blister. Then leave the stripper to work.*
2   *When the paint is putty soft, begin stripping. Use a pliable; flat blade for flat areas, and a rounded shave-hook for curved moldings.*
3   *Clean the surface when the stripping is completed by washing it down with whatever solvent is appropriate to the stripper.*

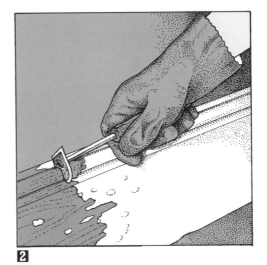

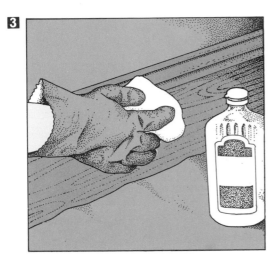

appropriate solvent for the stripper. After the surface has dried you can paint on it, first priming it because it will be absorbent.

BURNING-OFF

Burning-off is generally considered quicker than using chemicals, and can be cheaper. It is much the best way to strip oil-based paints. On thick build-ups it is certainly speedy; on stone or large areas of plaster that are heat-absorbent, not so — and you run the risk of cracking stone. This method is far less messy than using chemicals because the stripped paint is dry and it isn't necessary to cover anything except the area immediately beneath. For a given amount of money, you can strip a similar area with electricity or chemicals, but electricity will be quicker.

There are three types of burner generally available: the new hot-air strippers, blow-torches and the old blow-lamps.

Blow-lamps have been with us for generations and are usually powered by paraffin or gasoline. There's nothing wrong with them and if you've got one already there's no need to change it for one of the newer types, although they are more versatile. The shortcomings of blow-

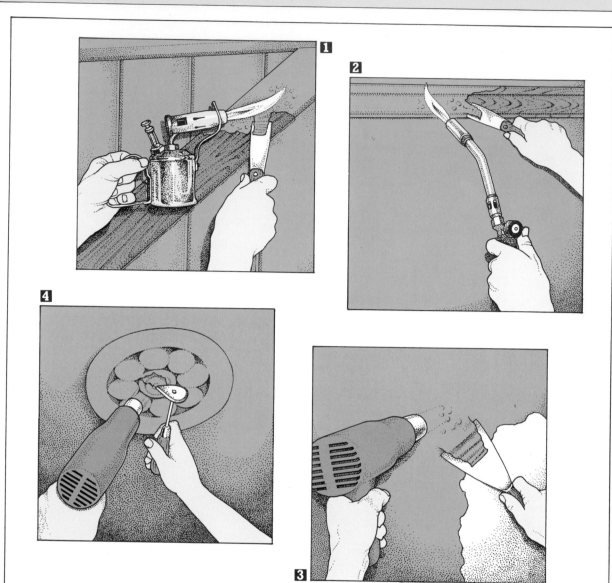

Burning off paint

1 Blow-lamps have a cool flame over a small area and the stripping blade should follow closely behind the flame, scraping as soon as the paint blisters.

2 Blow-torches are hotter and can be played over a larger area — about 1ft (30cm) square — with the blade following upward. They are very fast.

3 Heavy-duty electric hot-air blowers cover large areas with no fire risk, and are effective on oil or latex paints.

4 Little electric strippers with small heating filaments are very useful for close, clean work on precise details, but are far too slow for larger areas.

lamps are that they have to be constantly refilled, and must be warmed up each time before use; their pressure is variable and they've an irritating habit of suddenly blowing out or flaring. Although their flame is cooler than that of a blow-torch, you sometimes end up setting fire to the paint and burning the surface because of

holding the lamp too close to the surface.

Blow-torches are more expensive and therefore are usually hired. They are attached to gas cylinders and have variable nozzles, which means you can adjust the shape of the flame and vary its intensity. Because the flame is hotter than a blow-lamp's, it works faster. It

doesn't have to be reloaded and it doesn't blow out. Increased speed may compensate for the expense — it depends what you think your time is worth.

Electric hot-air blowers are the third type of burner. These aren't as new as they're made out to be, and there's a very definite difference between them and the very recent electric strippers that resemble hair-dryers. Hot-air blowers work by blowing air along a flexible hose and over a heater with a variable control in an insulated hand-piece. In many ways these are excellent. They avoid the obvious risk of a naked flame, and unless they are jammed up against wood for a long time they won't scorch it. They also work well on plaster and are excellent for detailed work like moldings or chair legs.

Be warned about the little electric strippers that resemble hair-dryers. They're adequate for very small, precise details but their heads consist of an electric filament like the one in a light-bulb, inside a guard nozzle. This means that they heat a tiny area of paint and are extremely slow. They're fine if you want something to do while planning your will, but try stripping a staircase with one and you'll get an idea of what eternity is all about.

■ **Tools** The tools for burning-off are a flat scraper as wide as the working area will allow, a narrower one, and a shave-hook. You also need to cover the floor. If you are working above bare floorboards, newspaper is quite adequate to catch the paint stripped by a hot-air blower but for the other types of burner aluminum foil is best, as hot slivers of paint will drop constantly. Paint which is melted and scraped off when hot dries and coagulates in hard lumps like plastic ashes and it can stick to rugs and linoleum like taffy. Also useful are a metal container for the shavings, and a bucket of water and some wet sacking in case you do set something alight.

■ **Application** Always work from the bottom up while burning off paint, as the rising heat softens the paint above and makes the work progressively quicker. Aim to heat about a square foot (30 sq cm) at a time but never keep the heat source in one spot — keep it moving constantly to avoid melting and scorching. The aim is to blister the paint, not to melt it into a sticky coagulation that clings to the tools and makes scraping like stirring Christmas pudding. As you move the burner or blower upward, use the scraper behind it, keeping the blade as flat as you can over the surface, so that you won't gouge grooves in it. On wood you can tilt the tool more than on plaster, but not at an angle of more than 15° from the surface, or you'll dig into the grain. Go with the grain if you're stripping wood, *never*

across or you'll tear the grain. Move from side to side on plaster. It may sound a bit like trying to conduct an orchestra and signal to traffic at the same time but, in fact, once the knife starts to glide up following the torch it becomes a rhythmical motion and the only real danger is losing concentration. Try to keep to the one-square-foot-at-a-time format, because paint that is heated and then not removed coagulates and is harder to get off than before.

PREPARING A PARTIALLY STRIPPED WALL IN GOOD CONDITION

If you have stripped defective paint from a part of a wall and find the rest of the surface quite sound, there's no need to continue removing it. If you find any defects in the plaster where you have stripped the paint, you should fill them. There are a large number of ready-mixed or powder composition fillers available from decorators' and hardware shops, expressly designed for this purpose, and many of them — especially the plastic filler — can be tinted with paint or stainers. If you have decided on the final color scheme, it is useful to color the filter accordingly. Fillers tend to be very pale, so if you are going to put dark paint over them it's better to darken them. Once you have filled the area concerned, you should sand it down with a very fine sandpaper, using a gentle circular motion, because paint will show up any ridging mark. If you intend to paint the wall with a water-based paint or latex, you should give the *whole wall* surface a coat of primer, otherwise paints like vinyl latex will always show up a filled area as a slightly contrasting patch. If the rest of the wall has gloss paint on it, touch up the filled area with oil-based paint diluted half-and-half with mineral spirits. Allow this to dry, and sand down the edges of the old paint, fill it again and retouch it, and let it dry before you apply the first coat of the new finish. That way you won't get patches under the new surface.

NEW PLASTER

The first thing to be sure of when painting plaster is that it is dry and flat. Of course it should be both, but you can't guarantee that until you look closely (and looking can be full of surprises). Dampness in plaster

Safety when Using Chemicals and Burners

When using chemicals, always wear rubber gloves; they can feel awkward, but are not nearly as uncomfortable as chemical splashes on the skin. If you are working overhead, wear goggles or old polaroid sunglasses, whether using chemicals or flame, and preferably cover your face, too. Use a damp cloth or a proper mask to protect your face from flame, although paper may be an adequate mask near chemicals — provided that you can breathe.

With burners, have some wet sacking handy, a stiff-bristled broom, and a metal container for hot peelings. A bucket of water is a good idea, too, with both chemicals and burners. Switch a flame off when you're not using it; if you pause for a short time, turn it right down but with a *visible* flame and stand it well away from any flammable surface.

Preparing a partially stripped wall in good condition

1 Test the drying time of the filler, and mix no more than you can use in a given time. If a dark color is to be painted over it, tint the filler while mixing.

2 Apply the filler with a pliable, broad-bladed knife or spatula.

3 When dry, sand the filled area with very fine abrasive paper wrapped over a wood or cork block.

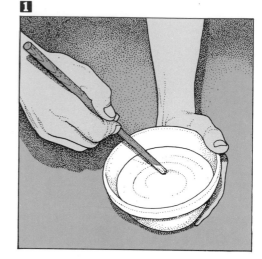

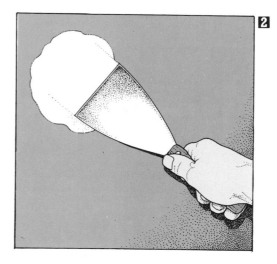

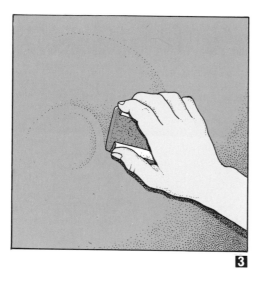

will give it a slightly piebald effect; the other main concern is that salts and acids may come to the surface under the paint, causing bubbles and cracks. If salts and acids come through water-based latex paint it doesn't really matter, because these paints are porous and the acid and salt appear like fluffy ash and can be brushed off. But if you intend to put on oil-based paint like gloss, flat-oil or egg-shell you should put an alkyd-resistant primer on first; this will seal off the acids and salts and stop them making bubbles and "mole hills".

As far as the regularity or flatness is concerned, good plaster has a smooth, sheeny surface like cold silk. If it's new it isn't really very hard as yet and it may have little ribs and freckles on it, no bigger than a small mole on your hand. You won't necessarily see these as you look straight at the wall, but paint will reveal them if you don't remove them because they catch the light; they can cause little tail-backs in the paint like tiny tadpoles. To find them, put your face against the plaster and look along the wall towards a light source and, if they're there, you'll see them. To get rid of them, use a 4in or 5in (10cm or 12.5cm) paint-stripping blade. Slide it flat along the wall and they will pop off: never,

Removing Lacquer and Gloss

Gloss paint comes off quite easily either with ordinary chemical strippers or with heat, although flames can make it rather sticky. Lacquer is trickier, and needs specific lacquer removers, usually in two or three applications, applied with a brush. *Do not use heat.* A cabinet-makers' scraper is ideal for removing softened lacquer, and a firm sponge is useful to finish off. Read the manufacturers' instructions carefully, as removers vary.

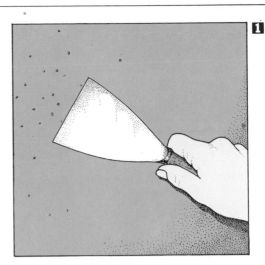

Painting on new plaster
1. Pop off any nipples left by the plasterer's trowel or any small flecks stuck to the surface with a broad, flat spatula, keeping the blade as flat to the surface as possible.
2. Fill any cracks and allow them to dry. You can sand the surface of the filler with *very fine abrasive paper, but not the plaster.* Then prime it.
3. After removing any fine nipples and filling cracks, you can wash the surface down with water and a cloth, or a very soft brush.

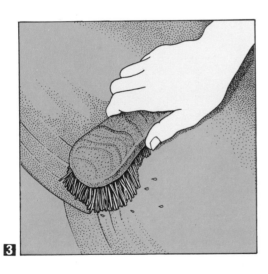

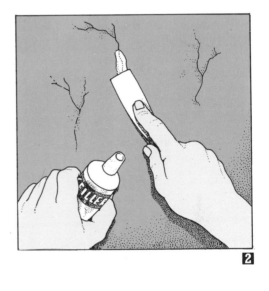

never tilt the blade more than a few degrees from the wall or you'll risk digging into the plaster. Absolutely *never* sand plaster that's new and fresh. Even the finest sandpaper will leave little wheeling scratches that will show through three layers of paint, and you'll only have to refill them and paint it all over again. It just isn't worth it.

If you are going to paint plaster with any water-based paint, such as latex, you should give the whole surface a coat of paint mixed half-and-half with water, to seal it. It will look horrible, although it's poetically called a mist or fog coat by many decorators. Its function is to make a

bridge or key for the top layer of paint, as much modern paint tends to lie on the surface rather than permeate the plaster, so don't worry about the mottled effect — it's supposed to look like that. This mist coat seals the plaster and stops the top coats going patchy. At this stage, you should fill any small cracks that might have appeared. Use the all-purpose, vinyl-based fillers from decorators' shops that are designed for this. Allow them to dry. You can sand off the filled cracks using a very fine sandpaper once you've sealed the surface around them. Then touch them up with the same half-and-half mixture. If the cracks

open a little, just repeat the procedure again, touch it up, let it dry, and you can then think about the finishing coats.

NEW OR NEWLY STRIPPED WOOD

New wood, which has never been painted, should be sanded down all over to remove any small fibrous splinters and also to key the paint. What you want is a surface that feels like smooth peach skin, is really minutely ruffled but has no rough areas that will show up on the paint surface. A fine abrasive paper or

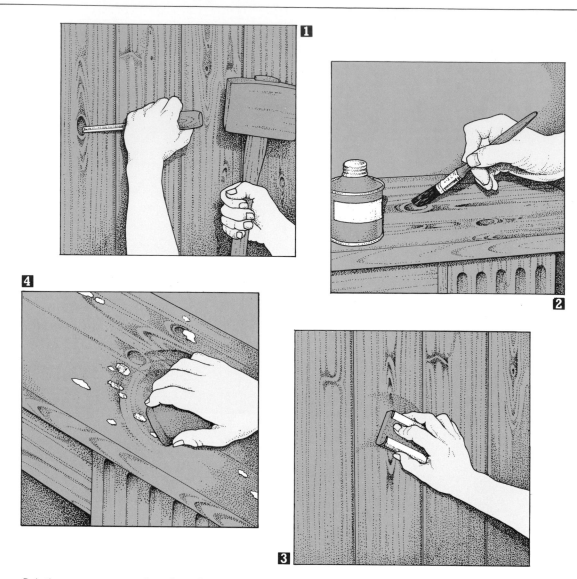

Painting on new or newly stripped wood

1 You can knock out any knots that may cause resin bleeding with a chisel, like coring an apple, and then fill in the holes with plastic filler.

2 Alternatively, give the knots and any resinous areas a coat of good knot sealer; this prevents resin bleeding out.

3 All new wood should be sanded down overall before sealing and priming.

4 Plastic filler is suitable for grooves, cracks and empty knot-holes. Filled sections should always be sanded flush and then sealed before painting.

fine steel wool is quite adequate for this on any planed wood. Most softwoods, especially pine, tend to have resinous knots that can "bleed" resin rather badly — it looks like corn syrup. If this isn't treated before you paint on the wood there will be sad streaks, like nicotine stains, fretting the paint. There are two things you can do about knots:

you can remove them with a chisel, or seal them. On boards you can often knock the knot straight out, as it has no lateral strength, using a chisel and a wooden mallet. If the knot gives downward it should drop out if not, you'll have to chisel it out like coring an apple. Then you can fill the hole with wood filler. You can paint directly on top of many wood fillers

and the plastic fillers and plastic woods can all be primed, in any case. Once you have filled the knot, let it dry and sand it flush with fine sandpaper. You may have to use a number of layers of filler to avoid sinkage in the surface.

If you don't want to try removing the knot, use either an aluminum primer, or knot sealer. Knot sealer is

Preparing Glass for Painting

Glass should always be washed before painting. It often has transparent grease and dirt on it, too small to see. Ordinary window cleaners are adequate to remove most of this, and glass can be scraped with a flat, blunt spatula. Dry the glass with a soft cloth. Remember that acids can etch glass, so choose your cleaner with care. Oil paints are semi-transparent on glass and need no underpainting; nor do latexes. Special glass-paints need a clean, acid-free surface.

made of pure shellac in wood alcohol. If you use it in a warm room you may feel drunk, and that is not just an illusion — the fumes are intoxicating. Ensure good ventilation. Use two thin coats over the knots and about an inch (2.5cm) around them. This will stop bleeding.

You must always prime bare wood before you paint on it. Use an aluminum primer on doors and window-frames, as they often come into contact with damp, and especially where wood abuts stone or brickwork. Aluminum primers are also best on any wood with resinous grain — that's an orangey grain that's hard and seems to stand above the softer wood around it. Otherwise, on all bare wood, including furniture, you can use a good, lead-free primer. It is worth stressing here that cheap paint is a waste of money and effort. Always buy the best paint you can afford, and that certainly goes for primers, too. Combined primer/undercoats have now appeared on the market and they save a lot of time as they make one process unnecessary.

The consistency of the priming coat depends on how absorbent the wood will be. Softwoods — for example, white woods — are very absorbent, and it's useful to add mineral spirits to help the primer to seal the surface. Hardwoods are much less absorbent, and for them the primer needs to be used thick.

On wood that has been stripped but still has traces of paint adhering to it, you have two options for preparation. One is to wet it and then rub it down with a waterproof abrasive paper or pumice stone. Then you should let it dry. Sometimes the grain gets puffy and swells because of the water. This is less common on wood that's been stripped by burning, because the wood has been hardened by the heat and is less absorbent than other bare wood; if it does happen, give the wood a sandpaper rubbing before you prime it. Alternatively, you can use linseed oil and mineral spirits to clean it. Mix one part of the raw linseed oil to three parts of mineral spirits, and rub it in with a pumice stone or self-lubricating paper. Then wash the wood with mineral spirits, rubbing it over with a lint-free cloth. It used to be possible to get painters' tack-rags for this type of job; they were called tack-rags because they were tacky, and picked up all the dust and gritty particles that always manage to get

onto surfaces like this. Ironically, as they were so useful, it's now almost impossible to get them; but you can make a fairly effective substitute by cutting a piece of old sheet or shirt — preferably white — and giving it a thorough soaking in warm water, then wringing it out and spreading it flat on a non-absorbent surface. Scatter turpentine over it as evenly as possible, then wring it to get the turps to flow through its fibers evenly. Open it out flat and sprinkle a dessert-spoonful of boat varnish (for a shirt-sized cloth) over it. Wring it out again, to spread the varnish all through it, and then hang it up for about 40 minutes. Fold the cloth up into a pad and it will take up just about everything dusty or gritty on the wood surface. This tack-rag will last for many months, provided it is stored somewhere airtight; you can re-treat it if it begins to dry out. Always shake it out afer use, or you'll put the muck back on next time.

VARNISHED AND WAXED WOOD

It isn't true that you can't paint over varnish. The glaze technique of oil-painting on gesso or canvas is based on exactly that, but it's not possible to over-paint varnish with water-based paints on plaster or woodwork.

If, for instance, you have a painted surface with crazed varnish (crackling, such as you see on old, glazed tableware), it's quite possible to apply oil-based gloss over this to maintain the cracked effect. You can apply two coats if you want the paint to be more opaque; one, if you want a glossier effect.

If you do decide to strip the surface, the quickest way of removing varnish, dirt and wax is to use a chemical paint and varnish remover. The method is the same as for removing paint (see above), but in place of the shave-hook it's better to use a cabinet-makers' steel scraper.

If you wish to leave varnish on but remove wax, or remove older, softer polish without using the strong chemical paint/varnish removers, you can apply a mildly abrasive brass polish. This will remove wax and leave varnish intact, if you use a cabinet-makers' scraper with a light pressure. To remove the varnish as well, use the scraper more heavily. For large areas covered with wax, use mineral spirits and steel wool or abrasive paper. If any residue still adheres, you can take it off with steel

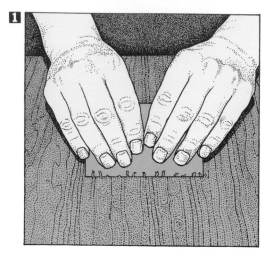

Removal of varnish and wax from wood

1 Use a cabinet-makers' scraper in conjunction with chemical stripper to remove wax and varnish. To strip wax and leave varnish, give the surface a coat of brass polish and run the scraper over with only moderate pressure.

2 To remove wax and leave varnish over large areas, rub mineral spirits over the whole area with a brush, cloth, steel wool or abrasive paper. In all cases, wash the surface down afterward with warm, soapy water and allow to dry.

wool and benzine. In any case, you should wash the surface down afterward with warm, soapy water, rinse it thoroughly and allow it to dry.

Shellac and lacquers could, until recently, be removed only with shellac and lacquer thinners, painted on with an old brush — as they destroy the bristles — but a new solvent has become available that will remove both. It won't, however, remove polyurethane varnish — that hard, glassy, general-purpose varnish that comes in matt, silk and hard gloss and seems to be an essential feature of the stripped pine furniture so popular in recent years. Polyurethane can be taken off with chemical paint/varnish remover, applied as above.

PAPER

It's a very common practice for people to slap paint quickly over old wallpaper, but this is usually a mistake. Firstly, once you paint over the paper the paint hardens and makes the paper twice as difficult to get off, should you want to change it. Paint makes wallpaper sticky, ragged and stringy when removed, because you have to break the paint film as well as the paper. Always

remember that paint has weight and weight is pulled downward; so if the paper is not keyed properly the paint will make it sag in places. Also, new paint causes old wallpaper to bleed dye into the paint, so that you get patches that look like distressed leather. The paint also betrays any roughness in the paper surface, and if the paper is fluffy or ragged, you'll get a hairy effect a little like old burlap. All the seams in the paper show up, and the effect can be like an old bed-sheet pinned to the wall after a dose of starch.

Unless the plaster surface underneath is really very dubious and you think it'll be more trouble than it's worth to remove it, you should take paper off a wall before painting. There are very few exceptions to this. One is if the paper is on gypsum block — often found in buildings that have been converted into apartments and have rooms or partitions; in this case, you'll have to leave the paper, as it won't come off gypsum block anyway. Also, you may want to retain the texture of a paper. If that is the case, the paper really must be firmly keyed to the wall. If there are any parts that aren't, take them off, and fill the area with vinyl-based filler. Don't on any account size old wallpaper — that's absolutely disastrous: the watery

Shading a Ceiling

This can be done from the center outward, or across diagonally. If you are working from the center, mix the center tone and those at the corners first, then the intermediate tones. One person should do the center and another the corners, simultaneously; the intermediate tones should then be added by one person, and blended by the other. Oil-based paint is best; latex is unsuitable, as it dries so quickly. When shading diagonally, use the method recommended for walls.

Stripping wallpaper

*1   Sometimes wallpaper will come off dry
with a knife. If so, work upward from the
bottom, gently and steadily, raising the
paper as you insert the knife.*

*2   If the paper will not come off dry, soak the
whole wall with very hot water. Start at the
top, as water runs down and makes the job
of saturation quicker.*

solution makes it swell and bubble
and wrinkle and, even if these
blemishes settle back and shrink
when dry, there will be air bubbles
riddling the back of the paper and it
will come away from the wall when
the paint dries. This can also happen
if you put water-thinned, water-based
paint on old paper. Either prime the
paper or, better still, give it a coat of
thin, oil-based paint. The trouble is
that some wallpapers bleed color
into the paint, especially red and
mauve dyes, which look rather like a
bad attack of sunburn. The best thing
is to leave the surface for about four
days and see what happens. If
bleeding starts, give the whole wall
or room a coat of knotting diluted
with wood alcohol. After that,
you should cross-line the walls to
hide the joints. Cross-lining is just
what it sounds like — putting lining
paper across, rather than down, the
walls. In all other circumstances,
take the paper off if you intend to
paint the wall.

Sometimes wallpaper will come
off dry with a knife, especially the
thick embossed papers and pre-
pasted papers. Paper can also be
removed with a knife if the wall is
very damp or the plaster is crumbly,
or if the paper has been steamed off,
as it can be in or near a kitchen. It's
always worth seeing if it will come off
easily. Do this by putting a knife up
under a *bottom* corner and then
sliding it upward gently, raising the
paper and gradually working
upward, as if removing a huge
Band-Aid. Always work from the
bottom up: if you work from the
top, the paper will hang down over
you as it comes off.

In the majority of cases, however,
paper won't come off dry. The most
basic method for removing it — and
one of the least messy and most
effective — is very hot water, the
hotter the better. Get a big, flat, 5in
or 6in (12.5cm or 15cm) brush, and
soak the whole area — if it's a room,
the whole room — thoroughly,

several times. *Always do this from
the top down*: this way the water runs
down and helps soak the paper more
thoroughly. Then, using a large flat
knife — a spatula is most effective
but you can use a large kitchen knife
— work downward but with
horizontal strokes; paper comes off
more easily that way because the
resistance of the width of the roll is
less than that of its length. Powdered
wallpaper strippers can be added to
the water, but they don't make much
difference to the results.

Most thick build-ups of paper
come off better by steaming than by
hot water alone. You can get away
with using a bevy of electric kettles
in a small room, if you have enough
of them; line them up along the wall
to create a small sauna and work
from the bottom up, because heat
rises. Over large areas of thick,
intractable paper, getting a steam
stripper is probably the best policy.
Steamers work on almost all papers,
and you can hire them. They have a

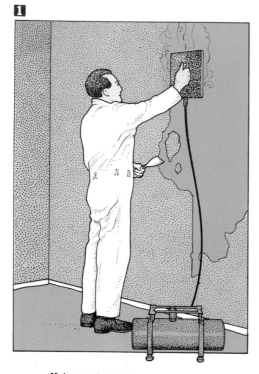

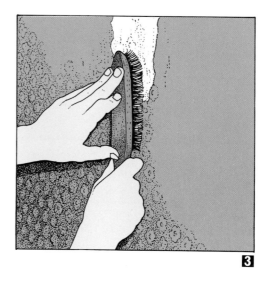

Using a steamer

1 *Hold the steam plate to the paper for about half a minute but no longer, as prolonged use may cause wall plaster to bulge. Work from the bottom up, as heat rises.*

2 *Score embossed paper with a sharp knife to allow the steam to penetrate.*

3 *Scrub thick papers before steaming, as this fractures the fibers and causes gaps for the heat to penetrate; after steaming, the paper will often come away by sharp brushing.*

perforated plate like the rose on a watering-can, at one end of a long hose; the steam is passed up the hose from a tank of water heated by electricity or by a gas cylinder. You hold the steam plate against the paper for about half a minute. This softens the paste and allows the heat to travel upward. Don't hold it longer than that in one spot or you'll run the risk of bulging the plaster. Run the knife upward and across from the bottom of the wall. The only paper that can resist a steam stripper is — predictably — the varnished paper of eighty years ago, which still lurks under others in bedrooms and hallways. On good old-fashioned

principles, this paper was built to last and last it certainly does. It's probably under there because no one else could get it off before you. The best approach is to score it carefully — to let the steam penetrate — and then try a kettle nozzle right up against it. If it surrenders and your knife starts to lift it, well and good. If it doesn't, then it's going to resist to the bitter end, so don't waste your time trying to steam it off: you won't be able to. You will have to cover it with another medium.

Paper that has already been coated with latex or any other water-based paint will usually need steaming. Heavily embossed papers

may need it if they've been painted, but if you scrub at the raised areas with a wire brush they usually come away. Vinyls can often be lifted dry by the bottom corner of each panel but the backing paper often remains stuck to the wall — you will have to remove that by the methods described above, although if it's very flat and firmly stuck you can paint over it.

The most difficult of the modern papers to remove are the washable ones, because they have been waterproofed by the manufacturers to avoid their being steamed off inadvertently. They are also thin and filmy. As with varnished

Paint on Metal Furniture

Most domestic metal-framed furniture is tubular stainless steel and cannot be painted. However, some older, cast-iron garden furniture may have found its way inside the house, and provided any rust is first removed you can paint it. Prime it with metal primer / sealer, then apply two coats of oil-based gloss paint, *never* water-based. Wrought iron requires its own special paint, obtainable from ironmongers.

paper, you have to score them to let the steam penetrate; but unlike the varnished papers, they will then come off with conventional soaking and scraping.

Once you have removed any of these papers preparatory to painting, wash the wall down thoroughly to remove all the old paste or size and leave it to dry. Then sand it down with abrasive paper and dust it off. After that you can prime it.

FABRIC

If a wall or piece of furniture is covered in burlap, canvas, linen or jute, you can paint on it if you want its texture; there's no need to remove it. Artists paint on canvas as a matter of course. Expensive fabrics like silk cannot be sealed, so they are unsuitable for painting on. Silk will become soggy in patches and, if it dries out, will be brittle. Felt is also unsuitable for over-painting, as its texture is too hairy to take paint well.

If you want to use a water-based paint like latex, all you have to do is prime the fabric with a thinned coat or two of it. Then allow it to dry before putting on two coats of normal consistency. Before using oil-based paints you will need to apply a warm, weak solution of size, to penetrate the fabric weave and stop it going brittle. You should let this dry — you can tell when it is dry as it lightens in color and feels dry — before you put on a primer and then a finishing coat. The drying times may vary according to the fabric, but follow the drying times stated by the manufacturer and allow a bit over just to be on the safe side.

Canvas can take almost any type of paint, provided that it is properly primed. Only water-colors (as distinct from water-based acrylics) are inappropriate to it, as their texture is too delicate for its weave. Canvas on a wall or modern chair is usually pre-stretched; that is it is already taut. Any paint finish that is already on it can be painted straight over although, if it is a darker color than the intended top coat, it may need a preparatory coat. Fabric on soft furnishings must be stretched taut before painting.

CANVAS ON FURNISHINGS

New canvas, if it is to be applied to a furnishing such as a screen, needs to be primed and stretched — which actually means shrunk — onto a frame or stretcher. This wooden frame should be constructed so that it has enough lateral strength to resist the pull of the canvas, which is extremely strong (a yacht's tubular steel mast will buckle before its canvas splits) and, if the stretcher frame is larger than 3ft x 4ft (90cm x 120cm), it should have at least one lateral cross-piece at the center. A sheet of linen- or cotton-duck canvas should be cut by laying the frame on the sheet and using a large pair of scissors or a very sharp craft knife to cut round the frame, leaving at least 3in (7.5cm) all around to allow enough canvas to be bent over the edges of the frame. Then, beginning at the center point of each side in turn, and working up to the corners simultaneously (not doing just one side first), either staple or tack the canvas onto the stretcher. Use rust-proof, bayonet-type, flat-headed tacks if possible, pulling the cloth as tight as you can, and making sure that the weave isn't stretched across the frame diagonally, but is square.

The corners should be folded last, in the same manner as folding the corner of a sheet under a mattress, and then pinned or stapled. The canvas at this stage should be taut, so that it is quite level when the frame is raised, but not as tight as a drumskin. You should then prime the canvas with acrylic or oil-based primers, preferably with two coats. For this size of work, artists' primers are usually the best and the acrylic primers dry very speedily. Priming can be done with an ordinary decorators' brush, loading the bristles liberally. Do not scrub the surface when you do this, as that will unnecessarily ruffle the surface, but work the primer well in with a cross-hatching movement. Coat the fabric so that the little holes in the weave show little blobs of paint if you look at the back of the canvas. As the primer dries, the canvas will shrink until it becomes drum-tight. After about 24 hours you should sand it over with very fine abrasive paper, with a quick, light circular motion, and give it another coat of diluted primer.

METAL

The types of metal usually met with in domestic interiors are not easy to paint; most common are galvanized steel — window-frames and radiators — copper piping and chrome furniture.

Window-frames are rustproofed before fitting, and that makes the keying of paint very difficult on any of the older ones from the 1930s to the 60s and, with the exception of some recent models that allow for painting, they are basically greasy when new and never intended for paint. The most common approach to painting sound, unrusted, galvanized steel is to wash it in mineral spirits and then in mordant. You have to clean it in such a way that the undercoat you use will key to the metal, and mordant — a chemical etching solution — will do this. After the mordant is applied, the metal should be washed thoroughly with water and left to dry, with no loose material being left on it. There are very good zinc chromate primers available, which also protect metal against rust, but unfortunately a lot of undercoats don't key to them. The only way around this is to read the manufacturers' recommendations very carefully.

Any attempt to paint over rusted metal is absolute folly and doomed to failure. You have to remove any rust before you can attempt to paint the surface. It depends if the metal is lightly or heavily rusted. Light rust looks like a pocky powder on the surface, and against this chemical rust removers are adequate but they have to be cleaned off very thoroughly or they make the subsequent application of paint very difficult. Tedious as it is, there is no real substitute for chipping and scraping to remove heavy rust. There is no special combination of tools for this: any scrapers, blunt chisels, spatulas, knives, wire brushes, abrasive paper or steel wool can be used — preferably the scrapers first, then the brushes, then the wool. Just keep on going until you work down to a bare, clean surface. That's all there is to it — and you'll probably think that's quite enough. There's no quick way of doing it if you're going to do it properly. You will also need thick gloves: gardening gloves or motorcycle gauntlets are very effective skin-preservers.

Radiators are commonly pre-finished in enamel, but this can chip and become grubby or rusty. So, if necessary, strip the paint. The best method is to use a chemical stripper and then to clean the rust, if any, off the bare metal. It is best to get the radiator hand-hot if you intend to paint it, and then to prime it with zinc chromate while it's warm. Whatever you do, *don't* apply a water-based paint to a radiator, because it will rust. Always use an oil-based paint. There are now special heat-resistant paints that can be applied to warm, dry, clean metal — but it must be all of those things. It is worth noting that gloss paints do not conduct heat as well as flat oil-based paints, because of their surface skin. All colors will ultimately fade on a heated surface but darker colors alter more than light ones.

Copper and chrome should never be painted. Paint simply doesn't stick well and flakes when heated. Even if you rub copper down with emery cloth and mineral spirits, you can't rub it dry; you have to let it dry of its own accord because copper dust floats and then leaves blue-green speckles like lichen wherever it falls. Copper always gets its revenge for the outrage of being painted, but if you really feel you have to, prime it with zinc chromate and then apply gloss paint directly to that with no undercoat. It won't last any length of time but it will outlast other methods; the copper will always win in the end. As both copper and chrome are really very handsome metals if left unmolested, why not just polish them and enjoy them as they are?

TILES

Tiles, if they have never been painted before, need only be washed down with detergent to remove any grease and then left to dry. There are some excellent brands of paint on the market expressly intended for tiles, their only drawback being that the color range is somewhat limited, as they are intended primarily for flooring; however, you can always mix them. There is also a wide range of enamel paints — from those used on boats to those used on model aeroplanes — which work very well on tiles and don't need an undercoat. If the surface is already painted and in good condition, you can paint straight on it. Latex and water-based paints are not at all suitable for tiles, as they tend to flake. In any case, painting tiles with that type of paint rather defeats the point of having them.

To clean paint from tiles, if necessary, it's easier to use a chemical stripper in the usual way. Do not attempt to burn off old paint; you run the risk of cracking the tiles, and they will absorb the heat.

Removing Old Distemper

If distemper marks your fingers when rubbed, it will usually come off with a stiff brush. Whether or not it comes off to the touch, brush it vigorously with a hard brush, then soak the surface with warm water and scrub it hard, changing the water as soon as it becomes milky. Then wash it down thoroughly with a sponge, allow it to dry, and prime it before painting.

Right Filling the pits and inevitable cracks which occur in plaster (after some years) can take quite a while. When removed, wallpaper may leave an acne of freckles on the plaster beneath. Plaster shrinkage from baseboards may leave lines of little holes. To ignore any of these problems will mar the final painted finish. Corners, in particular, catch the light and should be pointed up sharply. The main thing to remember is that preparation takes longer than application of the finish, but it's worth doing properly. The paint will last far longer and look infinitely superior on a well-prepared surface, giving a far more professional appearance.

4

BROKEN COLOR

Using one color over another gives multifarious variety of pattern,
texture and hue. This is the essence of all broken color techniques.
Methods covered here include color washing, shading, sponging,
stippling, dragging, and combing, ragging and rag-rolling, and spattering.
Antiquing — using these techniques to imply age — is also covered.

Broken color means applying one or more colors in broken layers over a different-colored background. This approach makes the most of paint's great versatility, and the variety of effect is equalled by the simplicity and effectiveness of the method. The results almost always give a unique one-off finish and offer wide-ranging options to those who feel that plainly painted walls look bleak, but cannot find or afford the paper or pictures they might prefer.

Techniques of broken color date right back to the Ancient Egyptians and the same methods are practiced all over the world. Professional decorators, however, have taken care that their methods remain secret, as most are so simple to

apply that in many cases amateurs can use them just as effectively themselves. These techniques vary in visual texture and depth as well as color and they have a very practical advantage: they can disguise the superficial imperfections of a surface by utilizing them as part of the process.

All broken color effects divide into two basic types: those where you add paint and those where you remove it. Where you add paint, as in spattering or sponging, you can use a wider range of materials; for example, you might use water-based latex paint and flat-oil paint, egg-shell and glaze, all on the same area. On the other hand, with the subtractive techniques, such as combing or stippling, there is less scope because you need a slow-drying paint that will stay wet, and therefore workable, longer. This means that the quick-drying, water-based latexes are more difficult to use than oil-based paint but you can use them to achieve a different type of finish with the same method.

Because latexes don't need an undercoat, you can keep them workable longer by applying a coat of the more slow-drying, silk-finish paints first and then applying the latex on top. The latex will then soak into the area beneath more slowly and stay workable longer. Latex has a softer effect than oil-based paint if you use it for ragging, combing, dragging or stippling, producing a range of cloudy effects from

Above left This light ocher wash gives both warmth and lightness to a high room. While setting the basic color tone, it allows great freedom to the rest of the decor.

Below left The contrast of matt wash to sharp, white woodwork accents the proportions of the room and allows the furnishings a clean, mellow backdrop which accords with their quiet warmth and weight.

Above right Delicate blush pink allows the silk lightness of this frame a sense of weight, where patterning on the wall would overwhelm it.

Below right The proportions of a high stairwell are beautifully accented by the soft wash to the upper wall area, offsetting monotone photographs, while the form of the banister work is balanced by the off-white of the lower wall.

distressed leather to thin cotton. The oil-based paints give a sharper finish which, in combing and stippling, is rather more sophisticated. The general principle is to use flat, oil-based glaze over flat-oil paint, egg-shell or undercoat, and a water-based glaze or wash over latex. A latex ground needs no undercoat — although it's best to put two coats on new plaster — but always be sure that the ground coat is grease-free or the decorative top coat won't take properly. If you are going to use oil-based paint on new plaster, put a primer/sealer on, then an undercoat and a low-luster egg-shell as a ground coat. It's worth repeating here that three or four thinned coats are preferable to one thick one. It's essential to let the ground dry out

completely before glazing or applying another decorative top coat.

THINNING DENSITIES FOR BROKEN COLOR

In broken color, all the top coats of paint are thinned, which is an economic as well as an aesthetic advantage; you can cover a room or stairwell area with a third to a half the quantity of paint that would otherwise be required. Thinning is necessary because the texture of the finish comes from the broken nature of the color, while the subsequent glazes give it a more translucent, deeper effect.

■ **Flat-oil, undercoat or egg-shell** All these paints can be thinned up to 1:1 with mineral spirits. The addition of any further solvent gives greater translucency but the mineral spirits can separate from the paint overnight, so don't mix more than you can use in a day.

■ **Glazes** Glazes for oil-based paint give the color a fine, glowing translucency. Their consistency varies but you can thin them 1:1 with mineral spirits initially. Remember that the more solvent you add, the quicker the glaze will dry, so be careful.

■ **Latex** Latex can be thinned up to 3:1 parts water to paint, or 4:1 to make it translucent.

COLOR WASHING

Color washing means applying a coat of thinned, and sometimes translucent, paint over a white or colored ground. This can be done with oil- or water-based paint but the term is frequently associated with distempering. Most people think of distemper as a powdery, off-white wash applied to old wooden houses or the lower walls of tenement blocks. In fact, it is a highly attractive and versatile medium that is very easy to use, offering effects varying from glowing translucence to a chunky, rugged texture that can enhance what seemed to be a hopelessly uneven wall. Apply it thickly on a rough surface — say in ocher — and it can give a texture reminiscent of oatmeal; applied thinly over the same type of surface it can give a shimmer like that on a lighted cave wall. You can lend a warm glow to a cold, north-lit room by giving it a wash of rose madder,

Above *This distressed effect gives the walls the appearance of being made of leather or old baize. Newly stripped walls after the removal of paper may have a similar appearance but rarely the same textural evenness. This distressing in color wash enables you to control the aspects of the stripped effect which are harmonious, timeless and extraordinary.*

yellow ocher or pale red-gold — all colors that might overwhelm a room if applied neat .

■ **Materials** Distemper has the asset of being very cheap; it is probably the cheapest of all paint finishes. As a result of this, many shops find it unprofitable to market and therefore don't stock it. Fortunately, however, distemper is very easy to make. All you need is whiting, (which is composed largely of crushed chalk that has been powdered, washed and dried); glue size; a tinting agent such as artists' acrylic, powder color, artists' gouache or universal stainers; two buckets and access to plenty of water. Make the distemper by breaking up the whiting into small lumps, putting it in a clean bucket, pouring cold water over it and leaving it to soak for 30 minutes. Then pour off any excess water and beat the solution to a smooth, thick batter. Distemper lightens as it dries, so you should test the color first, before adding the size. Dissolve the stainer separately in cold water, because if you add it dry you'll get streaks of undissolved pigment; then stir it gradually into the whiting. Take it slowly — you can always add more pigment but you can't subtract it. Test the color on a piece of thick paper and dry it in the sun or before

a fire. Bear in mind that when you add the size, the color will darken slightly. When you have the color you want, add the size. This, too, should be mixed separately, according to the manufacturers' instructions. Use hot water and blend the size solution with the whiting while still warm. If by chance you've mixed the size and haven't been able to add it before it has set, heat it over a pan of water on the stove.

Adding size to distemper is a delicate affair because too much will make the distemper flake and too little will make the mixture too powdery to adhere to the wall when it dries. The safest rule of thumb is that the distemper/size mixture should have the consistency of standard latex paint, when it is to be used as a basic ground coat. For a wash coat, thin it to a milky consistency with water. In either case, only mix about as much as you can use in a couple of days; otherwise it will start to go off.

■ **Application** You can apply distemper over any sound, flat or gently undulating, dry, clean paint surface except old distemper, which will pick up on the brush and produce a patchy ground. You should wash off any old distemper thoroughly and then coat the surface with pure size or claircolle. Claircolle is a mixture of size and whiting which gives a white, uniform base ground for the first coat of tinted distemper.

Like most water-based paints, distemper dries very quickly and this is the main problem with applying it. To lengthen the drying time, some English theatrical scene-painters add treacle to it, about one dessert-spoonful to 1¾pt (1 liter), but you can use glycerine in the same amount if you can't get treacle. Close all the doors and windows and apply the distemper liberally and quickly, starting with the window side of the room and working inward. Direct your laying-off brush strokes toward the light; if you miss a portion of wall, it's better to touch this up with a sponge, as brush strokes are difficult to blend in once the surrounding area has started to dry. When you've finished, try to get as much air circulating in the room as possible by opening all the doors and windows. If you've just put on a ground coat, preparatory to a final wash coat, leave it to dry out for at least 24 hours.

OTHER METHODS OF COLOR WASHING

The method that comes closest to easel oil-painting is applying a wash of flat-oil, mixed in a proportion of 1:8 with mineral spirits, over an egg-shell finish. This gives an overall, even, nearly matt, translucent quality to the finish and great depth to the color. You have to work reasonably quickly, as the mineral spirits acts as a drying agent, but this method usually avoids any brush marks or hard edges.

With latex, you can obtain a very subtle 'distressed' texture by thinning the paint with water and using the brush strokes in a criss cross 'hatching'; this transforms the difficulty of latex finishes into an asset. You then over-glaze again on top of the distressed effect so that the ragged areas of brushwork show through. The most translucent of all washes consists of pure pigment and water, which needs to be mixed with a small amount of latex to give it some body; use about two table-spoonsful of paint per 1¾pt (1 liter) of water, and use artists' gouache as the main tinting medium. Gouache pigments are very powerful, so add a small amount first and more if necessary. If you are adding tinting pigment to flat-oil wash thinned with

mineral spirits, use artists' oils for tinting. In either case, always mix the tinting agent thoroughly in its appropriate solution first — mineral spirits for oils, water for gouache — before adding it to the thinned paint, as a speck of unmixed pigment can leave a long streak right across a wall. In all cases except flat-oil, washes should be applied to a flat, latex-painted surface. The surface should be grease-free, as wash will simply run off greasy patches. These can be found and removed by wiping the surface with a water-vinegar solution after washing it and then rinsing it thoroughly.

■ **Application** If you decide to apply the distressed effect with latex as mentioned above, apply the color wash liberally and loosely, brushing in all directions and either leaving areas of the base coat unbrushed or covering them only very lightly. Allow this application to dry for 24 hours, and then apply the second wash coat. Put this on thinly and evenly all over the surface, so that the distressed patches show through, their color being enriched by this second coat, while the undistressed areas will appear softer. If any wash starts to run, just work it in with a sponge or brush it vigorously. At the half-way stage the general effect can

Above A soft, moss-green color wash with a gloss varnish finish. Gloss paint is highly unsuitable for such a wall area and silk finishes do not possess the element of depth. Varnish offers not only protection but also a translucent glow that retains the color beneath it.

appear rather bedraggled, as if the wall were a half-molted snake skin, but don't get depressed — it should look like that. The third and fourth wash coats will merge in, and if you have chosen different colors for them, they will produce a series of veils of shifting, translucent color with a luminosity that can't be achieved with any other method. The distressed areas will be of more intense color, the other areas of a softer effect with an element of depth.

If you have not distressed the surface and apply a number of even washes you will, in effect, be applying a series of color filters to the wall, rather like translucent sheets of misty, colored glass, one on top of the next.

It's best to leave color-washed walls in their matt, rather unfinished-looking state. This is because when outlines of woodwork, such as doors and baseboards, have been sharpened up the broken color walls look more intentional and their panache and beauty are accentuated by the contrast. You can apply broken color to woodwork but it is less advisable. The reasons for this are hard to pinpoint when talking about hypothetical rooms but suffice

Above *Shading offers a delicacy and ambiguity of light to wall and ceiling areas of almost any size. Depending on whether the room is used throughout the day or only at certain times, the color shift can be used to achieve a dramatic effect in strong sunlight or bright artificial light or gentler effects in softer light. Remember that natural light shouldn't contradict the direction of the shading, or you'll lose the transition altogether.*

it to say that walls and woodwork both given the same treatment lose form and strength; at best they look cluttered and shapeless — at worst they have the vulgarity of a comic-opera palace.

If you want to varnish broken color, use a matt polyurethane varnish; it will still give a slight sheen but less than any other type.

SHADING

This involves blending transitions of color from light to dark across a surface. It's a process that can most

easily be done by two people, and it can be used to adjust not only the light effects but the proportions of a room very effectively. A very high ceiling can be visually lowered by beginning with a pale tone at the bottom of the walls, gradually darkening toward the top and finishing the ceiling in the deepest color. The reverse effect is achieved by reversing the process. Walls can be "softened" by giving them a pale tone in the center and deepening it out toward the edges. Shading can be done with tones of the same color — say, pale magnolia shading to golden buttermilk — or with harmonizing tones of different colors; that is, different colors of equal brightness. If you're using different colors, you must be careful in your choice and in arranging the harmonizing sequence. Be careful that, if you place two complementaries together and blend them, you don't get a murky gray at the transition point. If you have pale sky-blue at one end and pale pink at the other you'll get gray in the middle. So look at the color wheel for intermediary tones, such as lilac or mauve, and blend those between the blue and the pink. Don't over-do shifts of color. The danger with this technique is ending up with something that looks like a stage back-drop, without footlights to quell it or a curtain to blot it out. Fire-red blending into violet and deep blue may evoke a dazzling sunset, but it can also look disturbingly like the burning of Atlanta. Shading works best with light pastels.

SHADING WITH OIL-BASED PAINT

■ **Materials** As with easel-painting, soft gradations of color are most easily achieved with oil-based paint; flat-oil is therefore the best choice for this technique. However, manufacturers tend to supply it only through specialist outlets in a much-reduced color range (partly, one suspects, because it's so versatile that it reveals the shortcomings of all the other so-called convenience paints; for example, the thick, slimy, non-drip types that can't be mixed, can't be thinned, can't be glazed and can't therefore be seen as convenient once you've worked with truly versatile paint). A good substitute for the excellent but difficult-to-get flat-oils is undercoat, which can be bought anywhere. For shading, buy a large can in the palest of the toning colors or white

and just tint it yourself. Dissolve the pigment — artists' oils are the best — in a little mineral spirits and then mix the solution into the paint gradually, stirring carefully and thoroughly to avoid streaking. Add pigments little by little — you'll be surprised how strong they are and how quickly they change the color of undercoat — and thin the undercoat with mineral spirits, as this will help blending. *Don't* thin it more than 1:3 parts mineral spirits to paint, or the covering power will be inadequate.

■ **Tools** Ordinary paint brushes are necessary for the application of undercoat, base-coat and the initial

or rough blending of the top coat, but the blending itself should be done with stippling brushes. These are rather expensive, though; large sponges are a good substitute and can be strongly recommended.

■ **Ground coats** If you are using flat-oil as a ground, apply an undercoat first. If you are using undercoat as a substitute for the flat-oil, apply an untinted undercoat full-strength and then the lightest shade of the tinted undercoat that you've mixed. Apply as many coats of this as is necessary to get a solid finish. Whether using flat-oil or undercoat as a ground, the last coat can be thinned with a 1:2

Above *A softly sponged finish. There is a harmony between the firmly plush quality of the furnishing, the crispness of glass and the quiet solidity of stonework that is created by the peach-like texture of the walls.*

mixture of linseed oil and mineral spirits to give the paint enough gloss when dry to aid manipulation of the shading coat over it. If there is a big variation between the darkest and lightest of the shading colors roughly blend the last ground coat, too, to stop the paler parts of it showing through the top coats.

■ **Top-coat tones** About three shades of color are usually adequate for a wall in an average-sized room. Mix the largest and darkest first and then the middle tone or tones. For a big wall or a very tall room (not quite the same thing) you may need five shades; mix the lightest and middle tone first, then the darkest. The more tones you have, the more they need managing, and while you want them to stay workable and need them thin for blending, the more mineral spirits you add, the quicker they dry. So, if you have a number of tones over a large area, add about one part of linseed oil to eight parts of paint to slow the drying. If there are any small areas that have been missed, which you want to slow down, use the same proportion of oil glaze.

■ **Application** It's useful to have a guide-line on the wall approximately where you intend the areas of blending to occur. Do this with chalk: a length of twine rolled in chalk is very effective, and light blue chalk is best because it vanishes in paint. Using a spirit-level, measure a series of equal spaces along the top of the wall and a similar number of equal spaces at the sides. Attach a weight to the bottom of the twine, then suspend this from each mark along the top of the wall in turn, 'snapping' the twine taut against the wall; this will leave a vertical chalk line. Complete the grid with horizontal lines, either with a spirit-level or with the help of someone else to hold the other end of the twine, 'snapping off' against the wall as before. Decide in which grid squares you want the transitions of shade and mark them with chalk.

Always begin shading with the palest color and work steadily toward the darkest. Brush the lighter color into the next as evenly as possible with the ordinary brush, moving like this right across the wall. If you can, change to a clean brush about half-way across, to avoid carrying too much light color into the darker. Then go over the roughly blended areas with a stippling brush or sponge. The blending should not be done wholly with the stippler

because the transition will be too abrupt; again, either change to another stippler half-way across or, if you can't rise to two stipplers, clean the brush. Ideally, it's good to have one person begin stippling while the other person works ahead putting on the color with the ordinary brushes. But always make sure that the same person stipples because no two people's techniques are the same; if you alternate, the result will look uncoordinated.

SHADING WITH WATER-BASED PAINT

It isn't really advisable to attempt large-scale shading with water-based paint. The water tension causes bands of color to show like tide-marks and the paint dries by evaporation rather than by reaction — as it does in the case of oil. If you are going to attempt shading with latex, a coat of oil-based primer is useful as this makes the surface less porous; a silk-finish ground coat is also advisable, as it is less absorbent. If you are going to attempt shading on lining paper, soak it thoroughly. It must be very securely stuck to the wall, and even so you can expect it to blister while you are painting, as it will be so wet, but it should shrink back. In any event, when using latex on a wall or lining paper, keep the atmosphere as damp as possible. Put wet cloths on the floor over plastic sheets if you can, shut the doors and windows and dampen the wall. When you paint, apply narrow bands of color, starting with the lightest and adding more of the darker tones as you go, stippling each blend immediately and moving on to the next tone. It really is an asset to have two people working on this together, one stippling with a stippling brush or sponge while the other moves ahead, applying and doing the initial blending. Latexes are more successful for shading small areas than large ones; oil-based paint is really the best medium for this technique.

SPONGING

Of all broken color techniques, sponging is almost certainly the easiest and perhaps the most fun. It's very relaxing because there are so many things you don't have to worry about, like drying times, even glazes, keeping a wet edge going, keeping the environment damp or laying off

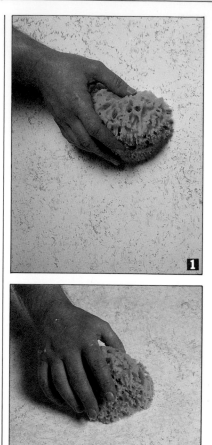

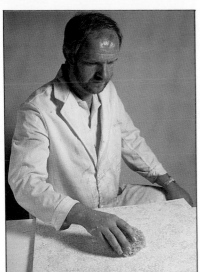

Sponging method

1. *Always begin with the darkest color, widely spaced, to give depth. Here, a watery bottle-green is used.*
2. *A pale beige-green is added, more closely spaced.*
3. *A pale creamy-gold over coffee brings the surface upward visually.*
4. *A light, random softening is given to the entire surface.*
5. *The result is a scattered, floating pattern you can look "into" not "onto".*

4

5

toward the light. You can forget the lot and dab away to your heart's content. As a bonus, if you use oil-based paint you can wipe it off with a rag soaked in mineral spirits, should the result be not quite what you wanted.

There are a variety of possible effects with this technique as there are three variables: the glaze medium, which can be translucent, opaque, matt or shiny; the sponge texture; and the choice of colors, which may often be a combination of three or four.

The variety of color depends on what you want. There are fresh, sharp combinations, such as black over emerald, emerald over tangerine, tangerine over white; there are various combinations of tones from the same color family, such as coffee over beige and caramel over that, or dusty blue, duck-egg blue and sky-blue; and highly sophisticated variants like electric blue over emerald or scarlet. If you use latex paint you can gain delicate, clouded combinations; for example, misty mauve and forest green sponged over sage, or dusty blue and pale olive clouded over pale lilac and dove gray. Whatever you do, remember that sponging in two colors works best with the lighter color on top, as this gives depth to the effect.

■ **Materials** Oil-based paints give a crisper texture to sponging, resembling densely crystalline stone; latexes give softer, cloudier mottles; if you want a translucent, marble-like finish, use an oil glaze. The best sponges to use are genuine marine sponges, because their texture is so varied, although ordinary cellular sponges are quite adequate — just watch out for their flat surface. Make sure you keep twisting them about with turns of the wrist between dabs to avoid getting any marking with the edge of the flat face that would make the pattern look too regular. Another way to avoid hard lines is to make an irregular tear through the sponge with the aid of a knife. Massage sponges have some interesting effects to offer and can be as stimulating with paint as they can be on the skin.

Besides sponges, you need a paint tray — the slanting type you'd use with a roller, with a paint reservoir at one end — clean rags, lint-free and undyed, and cartridge or lining paper on board.

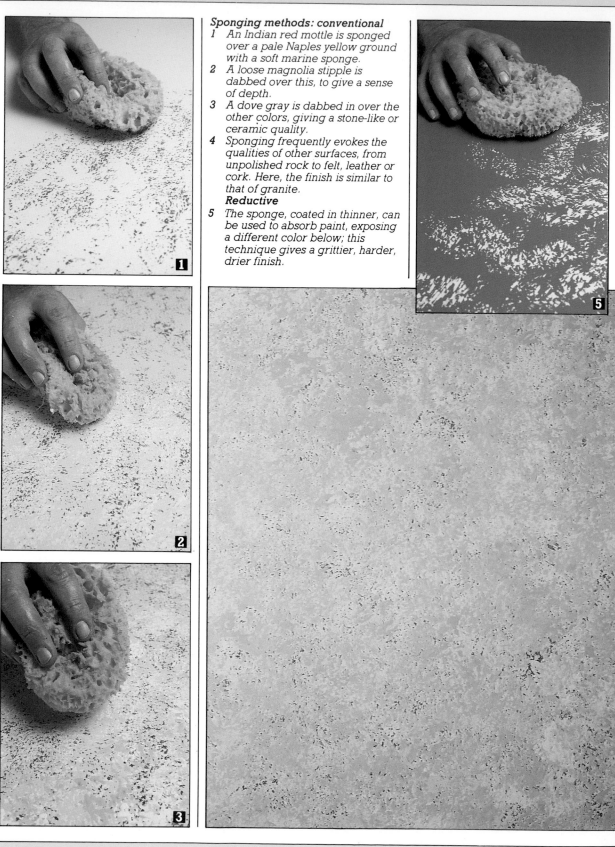

Sponging methods: conventional

1 An Indian red mottle is sponged over a pale Naples yellow ground with a soft marine sponge.
2 A loose magnolia stipple is dabbed over this, to give a sense of depth.
3 A dove gray is dabbed in over the other colors, giving a stone-like or ceramic quality.
4 Sponging frequently evokes the qualities of other surfaces, from unpolished rock to felt, leather or cork. Here, the finish is similar to that of granite.

Reductive

5 The sponge, coated in thinner, can be used to absorb paint, exposing a different color below; this technique gives a grittier, harder, drier finish.

■ **Application** If you use a real marine sponge — always preferable — it's best to soak it in mineral spirits first to soften it for use with oil-based paint, or in water if you're using latex. Let it expand again to its full size and then allow it to dry before using. Dab the sponge *lightly* into the paint tray and test it out on thick paper. Do this until you get a well-defined impression; then use the sponge on the surface. Test it on the paper each time before you apply it. This may sound laborious but it isn't: it just needs one or two dabs to check there's no excess, and then straight to the surface. If you are applying a color that is to have another color sponged over it, keep the prints even and well-spaced, wait for them to dry and then fill in between and over them with the other color. Try to avoid enervating repetition of pattern, by altering the position of the sponge in your hand. Wash the sponge regularly in mineral spirits or water but always wring it out thoroughly or you'll risk diluting the glaze or paint too much.

You can also produce a foggy or cloudy ground by sponging, which is an effect ideally achieved by two people. One lays on the base color and the other follows, sponging another color into it. When this dries, you can sponge on the top effects.

Cloudy effects made by sponging varnish over latex can be matt or gloss, unlike those achieved with color wash. Sponging varnish on top of the sharper, oil-based paints creates a surface like exotic polished stone, particularly suited to furniture.

STIPPLING

Stippling is the first cousin of sponging; it is really the same process, only using a sharper implement — the stippling brush. This is a large, flat-faced, soft-bristled brush which is dabbed on wet glaze and lifts just enough of it off for the background color to show through. The result is a fine, mottled texture sometimes compared to orange-peel. However, stippling brushes are extremely expensive and it is often advisable to find an alternative if you don't think you'll be using one often enough to justify the cost. If you are going to stipple a small area like a table, the best substitute is a soft shoe-brush; but if you intend to stipple, say, a whole stairwell wall — which, even in a two-

Above *Immovable or intrusive objects such as radiators can be camouflaged into an interior by careful sponging.*

Sponging Furniture

Either the very soft finishes offered by latexes or the very sharp, vividly colored mottles created with oils work best on furniture. It's advisable to be definite. Light, leggy furniture shouldn't be sponged except in oils, because soft effects just won't show up on it. Large furnishings — such as chests of drawers — work well with soft, marbled effects provided that they are protected with varnish, while table-tops and desks need a sharper, denser mottle, such as porphyry. The contrast achieved by sponging opening panels — such as drawers and cupboard fronts — while leaving the rest of the chest or cupboard unmottled, creates an interesting effect.

storey house, is quite a large area — you can get very tired with so small a tool and fatigue can make you careless, so that the stippling can become gashing and smearing. On a very large area, a sharp-bristled broom can be very effective. The long handle can be useful if you have to make a long reach; otherwise, remove it. Cut the broom bristles down with a saw or use a broom that's already worn down. You can stand back and use the broom from hip-height in a comfortable manner, you can reach a ceiling with ease and, if you are half-way up a ladder, you can just as easily reach down; it saves you climbing back and forth.

For very small areas — smaller than those where you'd happily use a shoe-brush — use a painter's dusting brush or a soft-bristled hairbrush. On an area of textured paint, large or small, you can use a rubber-tipped stippling brush, which is cheaper than the standard stippler but gives a rather coarse effect. All these tools produce a different texture but they all have one thing in common: a flat bristle surface, which is essential. Stippling won't work without it.

You don't have to use brushes at all. Alternatives include the attractive marine sponge with its softer stipple; a cellular sponge cut down the middle with its crisp, granular markings — the sponge should be rotated from side to side as you work; clean, undyed, screwed-up rags — similar to the method of rag-rolling; and even rollers.

In fact, rollers are the quickest method of stippling, but the hardest to control. If you use a roller, use the very coarsest you can get. Real or synthetic mohair rollers are the best; wool ones are good, too, but the smooth polystyrene ones won't work for stippling; they'll spread glaze all about in swirls like those you see if you close your eyes when you've got the 'flu. These tools all vary in finish but none gives the open, sharp texture of brush stippling.

The paint types for stippling can be just as varied as the tools. You can try stippling with matt paint over a shiny surface, or with low-luster over matt. Most traditional treatments consist of stippling a transparent or semi-transparent glaze over a white or light-colored ground. The clean, solid glaze color is broken by the tiny fleckings of the stippler to reveal specks of the ground color underneath. For example, a soft orange-gold over magnolia gives the velvet appearance of a lightly

Top *Linear moldings in unbroken color frame stippled flat areas in this pleasing window panel.*

Above *Fine stippling over an entire wall is time-consuming but creates an interesting visual texture.*

Stippling Table Tops

Stippling can be applied to give a loose, overall finish as on walls, or can form quite precise patterning. The only danger lies in applying too many colors. The geometric maze patterns used in formal gardens work well on table-tops. The variety of stippling textures augments the color variation between the areas, which can be divided by masking tape to prevent the stipple "straying".

powdered cheek. Salmon stippled over white looks like peach skin, coffee over cream like natural suede, electric blue over dove gray like shark-skin. Multicolored finishes give you the chance of blending a shifting mottle with no line of demarcation and with the variety of a cloudscape. It's quite possible to stipple either walls or woodwork but it's not a good idea to do both because, like sponging, stippling needs a contrasting finish to set it off. Stippling is suitable for use on furniture, but it is advisable to limit the number of colors, and to display the item against a plain surface.

■ **Materials** After applying your ground color with an ordinary brush, and selecting your stippling tool, you need pigment for tinting, a thinning agent, a couple of ordinary, flat paint brushes for applying the glaze and plenty of clean rags or paper for wiping the stippler and mopping up.

■ **Application** Stippling is very easy and simple, but more convenient and quicker with two people. Make sure that the ground color is dry, and then apply the glaze tone with an ordinary brush in a vertical band about 2ft (60cm) wide. Ideally, once this is applied, a second person should begin stippling the glaze while the first begins the second band. The stippler should leave about 3 – 6in (7.5 – 15cm) unstippled on the right-hand edge (when working from left to right); when the next band of glaze has been applied, go over the join and keep moving into the new band of glaze. Laying on and stippling take about the same time, so it is possible

to work without one person leaving the other behind. The glaze should be brushed out to a fine, even film and the stippling done by pressing the tool flat to the surface with a gentle but firm, decisive, dabbing motion — taking care to avoid skidding. Only one person should do the stippling, in order to achieve uniformity of touch.

Eventually, the stippling tool will get overloaded with glaze and should be cleaned by wiping its surface in mineral spirits. If an area is missed, it's easier to brush glaze onto the stippling tool's bristles and dab that on than to try to put more glaze on with a brush and then stipple it off. If you accidentally put two glaze coats on the same spot you can usually reduce them with a clean stipple tool; if the glaze is very stubborn and partly dry, moisten the area with mineral spirits and use a clean stippler.

If you use a roller, only one person should do the job. You have to keep the pressure even and avoid skidding. Otherwise, roller stippling is so quick it takes less time to stipple the glaze off than to put it on. You must be sure to keep the roller clean, though, by rolling it regularly on clean paper to take off any glaze build-up.

When you've finished, wash the tools in mineral spirits; rollers and sponges should be rubbed over paper and squeezed out in this solvent.

■ **Note:** *Be very careful if you are discarding rags on which solvent, paint and glaze have collected during cleaning. Let them dry out thoroughly first. Damp, screwed-up rags, steeped in this mixture of chemicals, are highly inflammable and if put into the black plastic sacks widely used for refuse collection can combust spontaneously, especially in warm sun. They are, in effect, potential gasoline bombs.*

DRAGGING AND COMBING

Dragging and combing have much to do with the representational technique of simulating wood grain, but dragging is a freer use of the basic graining technique and combing is a simplified stylization of it.

Dragging is basically dragging or coasting a dry brush over a wet glaze. Decorators most commonly use the brush vertically to achieve a

series of fine, variably spaced lines, which reveal the base color and leave the surface texture not unlike woven cotton, with pronounced vertical strands. The addition of a second glaze coat means that another series of dragged lines can be applied either vertically or horizontally, and a second set of verticals in a slightly different color can give the surface the appearance of raw silk. Verticals and horizontals together resemble a colored weave of heavy gauze; while applying the glaze in different directions, but dragging vertically, can achieve a shot-silk effect as the light strikes the glaze particles at different angles. Dragging usually creates a general effect of something other than paint.

The prime difference between combing and dragging is one of fineness. Dragging has a far softer, gentler appearance than combing, which is broader, starker and rather more dramatic in pattern. Both techniques can be used on walls but combing is more suitable than dragging for floors because of its heavier effect. Dragging is more suitable for woodwork, including furniture, as combing can look harsh — as if the grain has suffered some strange, woody rigor mortis.

Combing is very much what it sounds like: running a comb over wet glaze. This technique differs from dragging not only in the final finish but also in the range of tools that are suitable for it. Any hard,

Above *Dragging gives a clean, subtly aged quality, where other broken color might have become disjointed, and lends an air of elegant restraint to the potentially busy designs of carpet and furniture.*

Above *This finely dragged interior has the appearance of being hung with dull silk.*

Right *A flat-faced brush gives close, fine lines similar to the texture of woven fabric.*

Far right *A soft brush gives a broader, looser finish closer to the look of wood grain.*

comb-like object will do, from chipped stiff linoleum to a fine hair comb. Whatever tool you choose, the stroke is firm and continuous and the lines never have the feather-like quality given by a brush. At their very finest, using a hair comb, the lines resemble the old silver-point engraving technique, but they are always forthright and without the organic quality of brush dragging. As a result, the colors, too, can be starker. With the softness of dragging, very dramatic contrasts like lampblack on scarlet do not work because the fineness of stroke prevents the strong colors from showing clearly at a distance of more than about 1ft (30cm) and the effect can be brown and muddy. Dragging works well in pastels, such as sky-blue over soft emeralds, or earth colors, such as Indian red over ocher.

The prime difficulty many people encounter with dragging is keeping the hand steady to make a drag stroke as straight as possible. Often, the problem is trying too hard: if you hold the brush very tightly — which means it shakes — and then press it against the surface too hard; the bristles have too much spring in them and the whole effect soon has an uncontrollable wobble. The best way to hold the brush is by the haft, resting the handle across your hand as you would the shaft of a pen. Remaining relatively loose-wristed, simply rest the bristles on the surface and in one stroke coast the brush down the wall. Many people doubt their ability to make such a straight vertical stroke on a wall. There are two things you can do to help yourself. Either use a plumb-line suspended from the ceiling, about 1in (2.5cm) from the wall, and follow it down; or use a straight-edged board, with nails in it to hold it off the wall surface, and run the edge of your hand or the corner of the brush haft along it as you make the stroke. You can perform horizontal strokes in the same way, keeping the board true with a spirit-level. If for some reason there is no way you can make a single downward stroke, stop the downward motion between waist-level and the baseboard, as any unevenness will then be well below eye level; then brush upward to meet your first stroke, feathering the join lightly. Stagger the level of these joins, so that you don't get a noticeable ripple effect of joins along the wall.

As both dragging and combing

Above *Dragging on furniture can be highly effective. Following the general direction of grain keeps the feel of wood but allows great freedom of color and finish. Here, the dragged finish forms a background for a stenciled floral design.*

are suitable for walls, this method of working is suitable for both. For one-way dragging, however, you should ensure that the walls are smooth and even because, unlike many other broken color techniques, the use of vertical lines stresses any irregularities.

■ **Tools and materials** As with so many other specialist tools, dragging brushes are expensive and, unless you expect to use them often, it is reasonable to seek an alternative. Perhaps the best is a large, wide, paper-hanging brush. Jamb dusters are also very effective.

Combing is less demanding as far as tools are concerned. Graining combs are the official tool, of course, and although these are not nearly so expensive as dragging brushes, alternatives are much easier to come by and a lot of fun. Stiff hair combs, especially those with big plastic or steel teeth, are very effective, and you can bend them or break the teeth to give variety of stroke. Any number of comb-shaped variations can be invented by cutting V-shaped slots out of linoleum or plastic, which are also easy to clean.

Dragging and combing are usually easiest and most successful with an oil glaze over flat-oil paint, undercoat or egg-shell. This is because the glaze dries more slowly over their non-absorbent surfaces and because their tough skin stands up to the teeth of combs. An alternative dragging coat is a mixture of glaze and oil-based paint over flat-oil, which gives a more opaque finish. As a third option for either a dragging or combing coat, egg-shell thinned to near-transparency with mineral spirits is effective, although it

Ferning

This technique creates a pattern that resembles the leaf-forms of frost on a window pane; apply the technique to a desk, table-top, door or wall area. Work downward (toward you) so that each successive curving stroke is superimposed on the last and they all draw in toward the center. On a desk-top, the apex of the leaf should point toward the desk chair. A 3D effect is achieved by dragging the first strokes with a soft brush, then combing the later ones.

is a very quick-drying mixture and so less convenient for those unfamiliar with the technique. Oil-paint over undercoat gives the flattest of all finishes; here the stripes tend to merge together rather more, but this can often be most attractive.

You can drag with latex over an undercoat or egg-shell ground. This gives the softest of all combed effects, which can be very lovely, resembling very coarse parchment. It is difficult to drag or comb latex over a latex ground because it dries so fast on the porous surface and a latex ground won't stand up well to combs.

Whatever you use, it is always wise to test the tools, paints and their effect on paper before you start, and always keep the tools clean with the solvent appropriate to the glaze or paint.

■ **Walls** Dragging or combing a wall is most easily accomplished by two people. One should apply the glaze in a useful working band about 2ft (60cm) wide, while the other drags or combs steadily from top to bottom.

It is very important to keep the wet edge of the glaze "alive", especially if you are dragging horizontally, because if you let the glaze dry or become tacky, it is very hard to get an effective drag or comb stroke through it. If you then put more glaze over the top of the dried area and try to comb through that, you will get a patchy effect as the comb will remove the new glaze but leave the dark mark of the older glaze beneath.

It is best to keep the glaze brushed out to a fine, even film to

prevent it running. Always make the laying-off stroke downward, otherwise, unless you are putting on a number of glazes and drags to get a shot-silk look, the changes of direction will show through the tracks of the drag or combing as a wet-sacking effect; the wall will look slightly baggy and as if it is covered with wire mesh. It is, of course, always possible that you might like that effect.

When you are dragging, always be sure to wipe the glaze off your brush after each stroke or you'll just put it on again with the next and the definition of the stripes will soon become unpleasantly blurry. When you approach the base of the wall, lighten your stroke progressively as this will prevent a build-up of glaze from brush pressure. Go delicately around light sockets and switches, in the same way. Keep a rag soaked in mineral spirits handy to touch out any smudges. If you are planning two dragged coats, make sure that the first has dried out thoroughly before you add the second. If you use a water-based glaze, you must give the completed surface a coat of clear matt or semi-gloss varnish; this isn't essential for oil glazes, but it's always advisable.

■ **Woodwork** Dragging is better suited than combing to doors, baseboards and furniture. Combing on woodwork other than floors tends to look wrong, because it implies some mutation of the wood grain without actually resembling it enough to look intentional. Especially if you are using free colors — colors not natural to wood — combing tends to look incongruous. Dragging on wood can be done with smaller brushes than those needed for walls, but the method is basically the same. The essential difference is that you should follow the grain of the wood, even if that grain is not readily apparent under the base coat. Nothing looks more peculiar than woodwork that seems to have grain going in impossible directions, like a center-board in a door with hundreds of short, horizontal grains; to have cut a plank like that one would have to saw through a tree trunk 50ft (15m) thick — a Californian redwood at least — and such a plank would never hold together. Although you aren't simulating wood grain, remember that dragging on wood always implies a grain, whatever color you may use. On a door, follow the usual painting sequence of doing the panels in order and

5

6

Combing sequence

1. A soft brush lays on a loose paint gloss.
2. A coarse rubber comb straightens the lines.
3. A stiff, fine comb gives a kinked, herring-bone texture.
4. The same technique applied again, at right angles, creates a woven effect.
5. A fine rubber comb can be used to create a semi-circular fish-scale pattern, often seen on plaster.
6. A broad fish-scale pattern is achieved with a soft rubber comb.

Right *A cross-hatching effect is achieved with a soft dragging brush by crossing a wet paint glaze.*

Below *Dragging with a soft, flat brush gives a gently striated pattern.*

Bottom *Crossing a waving pattern with straight strokes, using a hard comb, evokes close-woven fabric — burlap or canvas.*

glaze/drag in one direction at a time, masking off each section and allowing it to dry before beginning the next. It helps to emphasize the joins between sections by a thin line in a neutral tone or slightly darker color than the glazing tone, either brushing this into a mark lightly scored with a knife, or filling in the score-mark with lead pencil. Other than this, the methods for dragging on wood are the same as for dragging on walls.

■ **Floors** Combing is better suited to floors than dragging, the soft, subtle texture of which is simply lost under foot and furniture. Combing can be executed on the bleached, primed wood of floor-boards and on floors of chipboard and hardboard. The parallel pattern of floor-boards, like the structure of a door, tends rather to dictate what finish is applied to them and most combing looks best following their direction and grain, whatever color combinations you may choose. These can range from mid-gray dragged over white, dark blue or mid-blue over pale gray, to scarlet over deep blue or magenta over golden ocher.

It is certainly easier to comb on the floor. Gone is the wobbler's bane of the vertical plunge to the baseboard that daunts the faint-hearted, and instead you can stop for coffee at the end of a floor-board without having to worry about the wet edge drying out. Hardboard and chipboard floors, with their smooth surfaces, offer extraordinary scope for patterning: effects range from those formed by diagonally wood-blocked floors to the curves and swirls one sees in marble. Combing can also produce the kind of parallel patterns seen on American Indian rugs or Spanish straw matting.

On floors, you must use flat-oil paint, undercoat or egg-shell. Latex will come off after only very little wear and is therefore unsuitable. As always, you must prime new wood and then put on a full-strength undercoat and at least two slightly thinned ground coats — preferably three. The combing coat should be a 1:3 mixture of mineral spirits to paint — thicker than you'd use on a wall. This is absolutely necessary to give enough body to the finish. You can use proper floor-paint, of course, but its dense texture is less suitable for the combing coat and the color range is very limited.

Remember that the oldest joke in decorating is the one about the person who paints himself into a

	COMBING AND DRAGGING			DRAGGING
Top coat	Oil paint	Egg-shell, thinned	Oil glaze	Latex
Ground coat	Undercoat	Flat-oil Undercoat Egg-shell	Flat-oil Undercoat Egg-shell	Egg-shell
Finish	Opaque, softly merging	Translucent	Sharp definition, semi-opaque	Soft
Drying time	Slow	Fast	Slow	Very fast

Above *Dragging on walls. The textures of floor covering and blinds are complemented by the fabric-like finish of these pastel-dragged walls. The direction of the dragging can accent the vertical or horizontal, too, whichever you wish to emphasize; this is virtually impossible to achieve with real fabric.*

corner of a room rather than out of the door — and it happens. Always work toward the door. If you are making a pattern based on any geometric design, copy the design on paper first and divide it into regular squares; then, using chalk, divide the floor into scaled-up squares of equal number. Patterns with straight edges, where one direction of combing ends and another begins — like a chessboard made of contrastingly patterned wood — should be painted with masking tape along these edges and the floor grid should be painted in a checkerboard manner, with the alternate squares filled in once the others are dry. Floor-boards are best done by coating and then combing about two at a time. Once the combing coat is dry, give it three coats of polyurethane varnish. This will ensure that the finish is tough enough to withstand the type of treatment floors inevitably receive and will not detract from the coloring.

RAGGING AND RAG-ROLLING

Rag-rolling is one of the most varied and dramatic effects in the field of broken color. It is the foundation of all marbling techniques and is best done in pastel colors because its startling patterning can otherwise dominate its surroundings. Rag-rolling differs from ragging in that the cloth used is rolled into a sausage of varying tightness and rolled lightly across a glazed surface. This creates a sense of movement that an unrolled rag does not give. The pattern

Left: Ragging and rag-rolling

1 In both techniques, paint glaze is first applied with a soft brush.
2 Ragging. A soft, overall, loose crumpling pattern is achieved with a soft cloth.
3 Rag-rolling. A more sinewy, creased pattern similar to marbling, with the cloth more bunched.
4 The soft finish of ragging.

Above A rag-rolled interior, reminiscent of a silky, polished basalt rock.

Right Ragging and dragging in the same tone can be most effective. Here, the ragging is contained in the frame of the wood-like dragging.

Above *Marble-like ragging suits an elegant context; here, it is in keeping with the picking out of the classical ceiling moldings.*

Top and far right *Camouflaging or assimilating inconvenient objects or blending sometimes awkwardly placed panelled surfaces is a particularly effective use of soft-toned ragging.*

Right *Marble-like rag-rolling of gray-blue paint glaze over off-white creates a visually interesting but relaxing finish.*

depends very much on the type of rag used; old sheeting, net curtaining, linen, lace, cheesecloth, jute and burlap are all suitable for rag-rolling, provided they are clean, lint-free and undyed. The crisper the fabric, the crisper the marking you get; the softer the cloth, the more various and subtle the effect. All marks left by cloth have a certain formality, though, and this gives a unity and rhythm to the surface finish that is otherwise seen only in some of the polished stones such as marble, or in the shifting yet harmonious sheen of crushed velvet.

As with combing and marbling, rag-rolling is more applicable to wall surfaces than to wood. There are a number of reasons for this. Primarily, the technique demands a reasonable amount of physical space to execute it and much woodwork is in places rather short of elbow-room. Secondly, the effect has often been emulated most inappropriately on latex enamels and, as a result of this misuse, ragging's application to wood can give the impression of latex enamel cut from panels that were designed for kitchen units — rather peculiar in a sitting room.

Also, applying ragging to woodwork as well as walls can deprive a room of form and contrast where it most needs it; the walls and woodwork may merge into a highly effective camouflage known as wave-mirror , used to hide aircraft flying low over water. A mottled vase on a rag-rolled table can vanish into a rag-rolled wall as effectively as two puffs of cigarette smoke into a morning mist.

On the other hand, rag-rolling can be used to make those banes of interiors, radiators, vanish with relative ease; the technique is well suited to their slab-sided form and they become one with a rag-rolled wall. On a wall, rag-rolling is really in its element and on the wide open spaces of ceilings it can be superb. It is particularly useful for visually defining surfaces that have an anonymous quality, softening angles, visually amalgamating oddly proportioned extrusions and alcoves and giving a sense of depth and sophistication to otherwise nondescript expanses.

Colors — especially those on latex enamel simulating rag-rolling — are usually muted because of the striking pattern produced, which might be too strident with louder or starker colors. Decorators tend to stick safely to light neutrals and pastels, but the potential variation is enormous. Linen, which makes sharp patterns with a soft-center effect within the stroke, can evoke crushed velvet when used with warm, wine-like colors — such as madder pink over magenta. This can be very oppressive if misused, but on small areas with soft light it can look sumptuous. If applied in light beige over cream, linen rolling can suggest a great sheet of crumpled paper or parchment.

Off-white grounds are highly versatile because you can put almost anything with them: off-white is a tint made with a touch of raw umber or raw sienna and looks superb with blue-grays such as French gray; gray-greens, caramelled pinks — pinks with a touch of ocher or burnt umber — and pale mustards. One of the most extraordinary effects is provided by rolling a white with a speck of black in it over an off-white ground, giving the effect of an old, damasked tablecloth. Then there is the white velvet effect of rolling off-white or a pale bluey-white over a stone gray.

Ragging — using a crumpled, unrolled cloth — has a more static

Ragging in soft pastel tones offers a stimulating yet delicate accent to finishes and textures beside it.
Left and top *Beside marble.*
Above *Beside wood.*

Right *Evenly applied ragging over walls and molded ceiling gives a unity and mellowness, where contrasting color on moldings would appear harsh, awkward and inappropriate.*

Ragging a Chest of Drawers

The surface should be well sanded and, if stripped, sealed first with primer and then with undercoat. The base coat and ragging color can be applied overall, if tonally very soft, without destroying the form and proportion of the object. Sharp ragging may well have a camouflage effect. Such furniture looks best with a satin-finish varnish.

appearance to its glaze stroke, and with thinned glaze the texture of the cloth is often more in evidence. Ragging with burlap, using ochers and pale yellows, leaves a distressed leather or chamois leather effect; with cheesecloth the surface takes on an appearance of soft wool. Ragged linen can evoke surfaces from creased parchment to mottled trout-skin. Old net curtaining, paradoxically, can make a wall look like dusty canvas or can produce an effect of great softness, especially with cloudy colors like pearly mauve or duck-egg green. Lace achieves a finish that evokes nothing in particular but gives a delicate, shifting patina of color.

■ **Tools and materials** Conventional paint brushes are necessary for the application of the ground coat and a large, soft brush for the application of the glaze. You will also need an appropriate solvent, plenty of pieces of your selected rag and a sheet of paper for testing effects. The consistency and type of paint most suitable for ragging are the same as those for sponging, which is the most similar technique.

■ **Application** As with sponging, ragging offers you a great deal of freedom and you don't need to worry much about drying times, as it is a very speedy process. The technique on walls is to spread glaze over a fairly large area — certainly larger than the usual 2ft (60cm) wide band — and make sure it is brushed out to a fine film, as evenly as possible; excessive thickness will mean rag-rolling movements will collect a heavy patina as they shift the surface and a blotching effect will result. Test the glaze and rag first on lining paper, going up and down and across the surface to practice the movement and to see what effects you get. If the effect is too heavy, either loosen the cloth and use a dabbing action or use a lightly rolling one, as if you were rolling flour into pastry. When you begin to work on the wall, keep a sponge ready in case you make too hard a stroke across the glaze. You can then touch up at the corners with the sponge or dab further glaze on with it and re-roll, using another cloth of the same type. A chamois leather is also very useful for this. You should have no difficulty with the glaze drying out before you've covered the area but when you put on the next band of glaze, blend the two sections together with very light strokes, finishing off always in the

same direction and avoiding dark, creasing smears. Leave a few inches of the preceding area unragged near to the join until you have applied glaze to the adjoining section, to avoid brushing glaze into an already ragged pattern, and then you can rag in over the join.

The exception to this method of application is latex paint. Because it dries so quickly, latex needs two people on the job, one painting and the other rolling. It's preferable to put the paint on in narrow bands and roll straight over them. One great asset of latex is its softness of finish,

and rag-rolled latex wash can be very lovely but it is a tricky business. Oil-based glaze is far easier to apply.

Change the rags regularly as you work, and try always to use a uniform pressure to give a consistent finish. The cloth itself will ensure variety but this should imitate the effect of broken clouds on a sunny day: all are different, but their weight appears the same. If you intend to apply a second ragging coat, you must let the first dry out thoroughly beforehand. It is also advisable to make this second coat a lighter

Above Rag-rolling in two tones gives a subtle sense of weight to the lower walls and raises the upper, giving a feeling of length, height and solidity.

Left The tight texture of this rag-rolling gives a greater sense of solidity than ragging, but remains highly versatile.

Below far left Rag-rolling as decoration.

Below left Rag-rolling as camouflage.

color than the first. As you approach woodwork it's useful to sponge the effect up to the edges, gradually softening it, as it is often awkward to maintain the rolling movement as you near the margin of the wood. It can be effective to add a further tinted glaze coat over a single ragged coat but not to rag that; this gives a sense of depth to the wall surface. Remember to lay the unragged coat off uniformly downward, to avoid a disruptive hatching effect.

■**Note:** *As with rags used for sponging, those soaked in glaze and mineral spirits should be disposed of very carefully to avoid spontaneous combustion, which can occur if they are left wet and screwed up, especially in sunlight.*

SPATTERING

Spattering is probably the most straightforward and, if successfully executed, one of the most attractive and stimulating of all broken color techniques. It is deceptively tricky to do well — probably because it seems so simple. It isn't just a matter of splashing paint; it really requires thinking in terms of a basic spray finish, and it needs quite a bit of practice before you try it out on woodwork or on a wall. It is a messy business, too, and you should cover any surface that you don't want given the same finish — including yourself. Of course, there are those who will tell you that if you do it properly you can work in evening dress and it is true that the key to this finish is control and forethought, not spontaneous anarchy.

Color is more important to this technique than any other consideration. Spattering can appear harsh or insipid, chaotic or measured, confused or harmonious; it all depends on the color and on the density of the spray. Provided that you are clear-headed, it pays to be bold. If you play safe and choose fading pastels, spattering can just vanish or resemble a failed attempt at sponging or stippling. These two techniques work well with pastels because they are denser; spattering looks best with bright, strong color over a large, pale expanse, or a dark or vivid color over a deep background on a very small area. On a large white wall, evenly placed sprays of tiny scarlet spots balanced with chrome yellow and deep bottle-

green work well. They make the wall look like a negative color plate of the stars: the neutral surface is defined by the spattering color, the size of the wall prevents that color becoming crowding and coarse. On a playroom wall, spattering can give a spontaneous panache that is hard to equal; or it can offer great sophistication — for example, walls with large, deep chocolate (burnt sienna) spatters on magnolia under a dark wooden ceiling.

Some sorts of furniture can be spattered, with discretion; for example, spattering stool-tops with dark chocolate brown over cream can give the effect of nougat slices of enormous size. Deep bottle-green or deep magenta are good base colors for furniture; tiny vermilion spots over deep bottle-green on a small table-top can give an effect not unlike the wing-scales of a rare meadow moth. Another stylish effect is to finish a table-top in the same colors as the ceiling, creating a mirror-like reflection.

■ **Tools and materials** For spattering you need a stiff brush. The best are stencilling brushes because they have short, flat bristles squared off abruptly; otherwise you can use an ordinary, coarse paint brush and chop the bristles off halfway along. As well as the relevant solvent and paint, you will also need plenty of paper, painted in the base color, for experimenting, and a straight-edged piece of wood.

■ **Application** Whatever paint you use for spattering, its consistency should not be denser than milk. First, practice mixing a small amount of the spattering colors until you get the proportions and consistency as you want them, and then practice your spattering technique. Don't hurry because the more you experiment, the wider the range of effects you'll get. There are two main methods of making a spatter and neither involves flicking your wrist. Nor do they require you to confront a wall in the manner of a Japanese expressionist and, with a loaded brush, make a cast as if shark fishing. If you want middle-sized spots (about ¼in/6mm), stand about 12in (30cm) back from your target and, with the brush loaded lightly at the tip, strike the brush against the straight edge of the piece of wood (a 3ft/90cm ruler is useful) with a short, sharp, chopping stroke. Don't lean close to the wood or you may contract instant German measles. If you want bigger spots, hold the

Top *Large, exhilarating spattering takes this radiator close to the realm of painted sculpture, a display object in its own right.*

Above *Spattering works well in conjunction with a wide variety of other patternings, especially on furniture.*

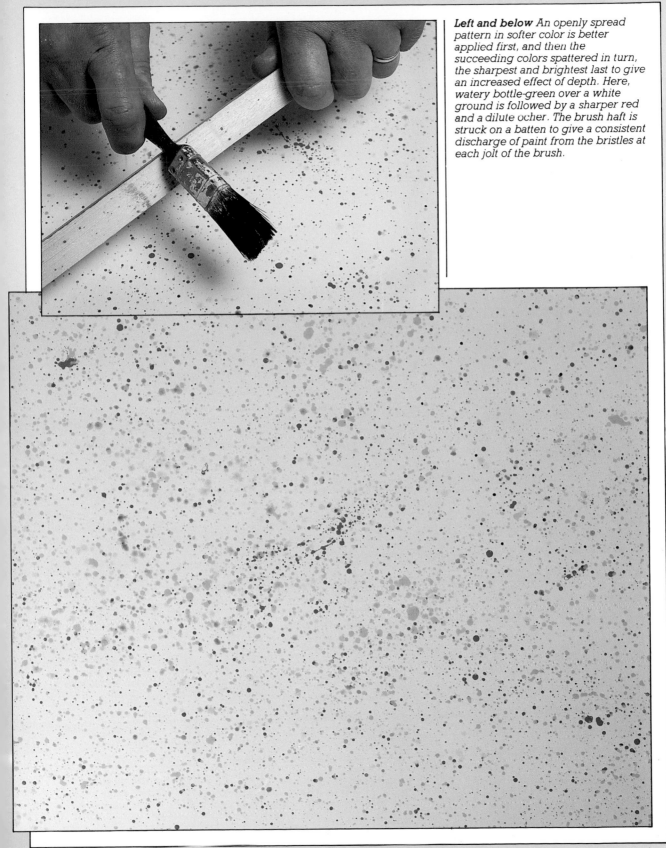

Left and below An openly spread pattern in softer color is better applied first, and then the succeeding colors spattered in turn, the sharpest and brightest last to give an increased effect of depth. Here, watery bottle-green over a white ground is followed by a sharper red and a dilute ocher. The brush haft is struck on a batten to give a consistent discharge of paint from the bristles at each jolt of the brush.

brush and wood closer to the surface. This is only a general rule and your spotting capabilities will be your own to develop; everyone differs. If you want masses of fine little spots, hold the brush about 6 – 9in (15 – 23cm) from the surface, and then run your index finger, a knife tip or, best of all, a fine comb along the bristles in a steady, one-directional saw stroke. With practice, you will be able to aim the flight of your droplets with surprising accuracy. Don't over-thin your paint — it should never be thinner than milk — because otherwise it may run down the wall. If you are putting another spray on top, it's a good idea to keep the spattering fairly open or you can get a crowded result. Once you are at ease with the technique you can really enjoy yourself, as spattering is one of those lovely techniques like sponging where you don't have to worry about drying times, laying off or any other finer details. All you have to do is make sure that one spatter layer is dry before you start another and be sure to protect the final result with varnish, especially if you have used a water-based paint.

ANTIQUING

To age an object all in a moment, look at it with half-closed eyes. Squint your eyes at a wall and its color dims; furnishings blur slightly; their sharp edges round. There are many techniques that will achieve this artificial ageing, with paint offering some of the most versatile and straightforward methods. Many ageing methods are really extensions of spattering, color washing and stippling, which have already been described so all that it is necessary to explain here is how they are used in a specific, softening way. The tools and equipment are the same, too, and it is simply a re-use of color and some care — and logic — that is needed when using these tools to simulate or stress age.

The first thing to remember in antiquing either a wall or a wooden surface is that age dims a shine and takes the edge off color. That doesn't mean it kills it — indeed, it may improve it considerably — but remember that what you should aim to achieve is an after-glow, a sunset on the surface, not a cloudy night. Using paint and paint glazes, there are three main techniques open to you: glazing and spattering on woodwork and color washing on walls.

WOODWORK

■ **Glazing** Paint glazes or oil glazes are among the most versatile antiquing approaches on woodwork, oil glazes offering the greater translucency. A 3:1 mixture of paint to matt varnish, tinting the paint first by mixing it with oil glaze, gives a high degree of transparency with a delicate, non-reflective finish. The chief difference between this and a straight oil glaze is that any glaze mixed with varnish cannot be rubbed or wiped when it is touch dry, while oil or paint glazes can. Therefore, for a wiping technique you have to use varnish-free glazes, either clear or tinted, and you can use them separately or together. If you use tinted and untinted varnish with a blending technique, lay the clear glaze onto the center of the area and then, with a stippling motion, brush the tinted glaze into it from the edges. On moldings it is easier to use a single, tinted glaze, apply it overall, and then rub the glaze off the raised areas with a cloth wrung out in mineral spirits, leaving the glaze gathered in the crevices. Then stipple it to kill any hard edges. If you use a tinted glaze over a flat area like a table, wipe from the center and blend with a stippling brush toward the edges. If you want a light effect, it is useful to allow the glaze to dry at least overnight and then rub it from the center with fine steel wool and blend it softly, leaving the edges darker. Then apply a second, thinned coat over that to give it a softer appearance.

The colors used in antiquing wood are similar to those of graining. In almost all cases they are the earth colors — burnt umber, burnt sienna and lampblack. A black surface is the only exception to this, where the antiquing colors should be dark brown.

■ **Spattering** Spattering is the key to evoking those "fly-specks" like freckles on an elderly face that typify so much old, polished woodwork. The most effective medium for this form of spattering is either artists' oil color or brown or black ink. In the case of ink, coat the surface first with shellac or varnish, and then apply the spatters in the usual way (*as described above*). With antiquing, the thing to note is *where* to spatter. Choose the places where the tiny gashes that collect dirt and old polish usually occur as these ultimately cause fly-specks — on the edges of drawers, the edges of level tops and surfaces, the tops and edges of raised turnings on banisters or furniture legs. Fly-specks do not usually gather on the flat, central planes, except here and there in a linear string, where any spattering should be much smudged and finely spread. Generally, don't make the spatters too regular, and vary their size. Blot or stipple the surface and, if you want to, ciss it with water over the ink or whisk the spatters with a soft, dry brush before the ink or paint sets. You can also reduce overdone areas with steel wool. Then varnish over the top of the spatter layer.

On wood surfaces, always bear in mind that a gloss varnish produces a harsh surface that looks totally wrong over antiquing. At most, you should apply a soft, satin sheen; use several coats of thinned clear matt or satin varnish or polyurethane and, if you wish, tint that to a mellow tone.

WALLS

■ **Color washing and glazing**
The main target in ageing a wall surface is to soften it, not darken the color. Walls that aren't flat and can't be made so — like many in older buildings — benefit from a matt color wash. A very thin wash should be used on them. The paler blues soften with a thin, raw umber wash. A gray wall can be warmed by raw or burnt umber either separately or together, touched with a spot of black. Green surfaces need a slightly deeper green that can be cooled down with raw umber or warmed up with burnt umber. White walls need a very thin wash of burnt umber, and the same goes for off-white, beiges, dusty pinks and yellows.

For level wall surfaces, oil glazes work very well. Their translucence, tinted delicately with the oil colors listed above, gives a patina of diffused softness that cannot be achieved solely with a paint wash.

Right Antiquing. Beams here have two coats of thinned, white paint, sanded off to show the wood. Walls have a white distemper and buff wash.

5

FANTASY DECORATION

Painted marble, porphyry and wood grain are among the oldest and
most beautiful of decorators' crafts, transforming walls, ceilings and
furniture alike. Painted tortoiseshell and bamboo offer an exotic
panache to smaller items, from banisters and coffee tables to clocks,
watch-cases and picture frames.

MARBLING

Marble has been imitated by painters for nearly 3,000 years: examples of it date from Middle Kingdom Egypt. The success and variety of painted marbles are inspired by the similarity between flowing qualities of paint and the nature of the marbles themselves. Indeed, the very term marbling is used to describe a particular type of paint effect that frequently has no pretensions to imitating polished stone; it means a whirling, mottled flow. Real marble — stone marble — did actually flow when it was forming, and the first thing to realize when simulating marble is that a sense of movement is vital.

Marble is created by heat and pressure on limestone, which crystallizes in white, black and a whole gamut of more brilliant colors — literally, all the colors of the rainbow. As a result of mineral structures running through the molten strata and cooling, the vein-like forms appear and the fragmenting of layers, like the laminates of plywood, allows other matter to fill the cracks before pressure welds the layers together as a solid stone again. But marble, crystallized out of liquid fire, possesses an elusive translucence, that of swirling clouds in a multicolored sky frozen in an instant, as captured in a photograph of a bright storm. Its veins are literally petrified flame — and so light shines through them. Look at marble and you seem to be looking at veils; below its surface you see shadows and ghosts of other movements. It is variably opaque and translucent — and so is paint.

There are as many techniques of marbling as there are types of marble but the movement of marble is always diagonal and, amid the swirling freedom of paint, you should try to remember this. Even so, there is no need to be slavishly representational in a domestic interior. There are examples of marbling by masters of the craft — and indeed in fine art, where its appearance is an intrinsic part of murals of the High Baroque — where, to the naked eye, the result is

absolutely indistinguishable from the real thing; but such skill demands observation, augmented by talent and founded on an intimate knowledge of both paint and stone; it is neither possible nor often desirable for the amateur to attempt this, and it is good to bear in mind that marbling means an effect — just as a book cover may be marbled — and not necessarily a photographic representation of the rock.

In today's interiors, marbling works well on walls and floors, baseboards and mantels and large, solid furniture.

With the exception of matching mantel and baseboard, it is best limited to any one of these areas in a given room. In most modern buildings, a room that has been wholly marbled may well look awesome and even rather absurd, because the structure and proportions of so many buildings are not suited to marble construction; however, marbling just one area within a room can look superb. Bear in mind that although you are not

necessarily copying the stone, you are extracting its essence — its presence. As you are implying the presence of stone, you should marble only those things that might logically be made of stone. For example, a marbled, three-legged milking stool is a visual contradiction in terms but a marbled table, provided it isn't of too light a construction, can work very well. Mantels are frequently real marble, window-frames never are; and only mythological tombs have marble doors, to be charmed open by flutes or pried open by emissaries of the underworld.

The flowing crystals of real marble contain all the colors of the spectrum, and the cool or sumptuous beauty of the stone may be a consequence of any combination of these colors. This means that there is no limitation to the colors you may choose for simulated marble, at least in theory. The most frequent combinations of real marbles are probably the most harmonious and to follow them is often wise. Otherwise

Cissing

Cissing is the technique of spattering droplets of a thinner onto paint to achieve a random, mottled effect. Normally, mineral spirits are scattered on oils, and water on latex, but the other way around can often be highly effective.

Method I Load a fairly stiff-bristled brush with thinner and strike the haft against a wooden batten to scatter the droplets. This is most effective when marbling, for implying a mottle between veins; in porphyry; and for tortoiseshell.

Method II If the thinner is blown in a fine spray through an artists' diffuser *(see page 106)* onto wood graining, the resulting pattern evokes the tiny satin-like pores found near knots and around the heartwood. As patterning in its own right, cissing works well on furniture, and is an excellent addition to sponging.

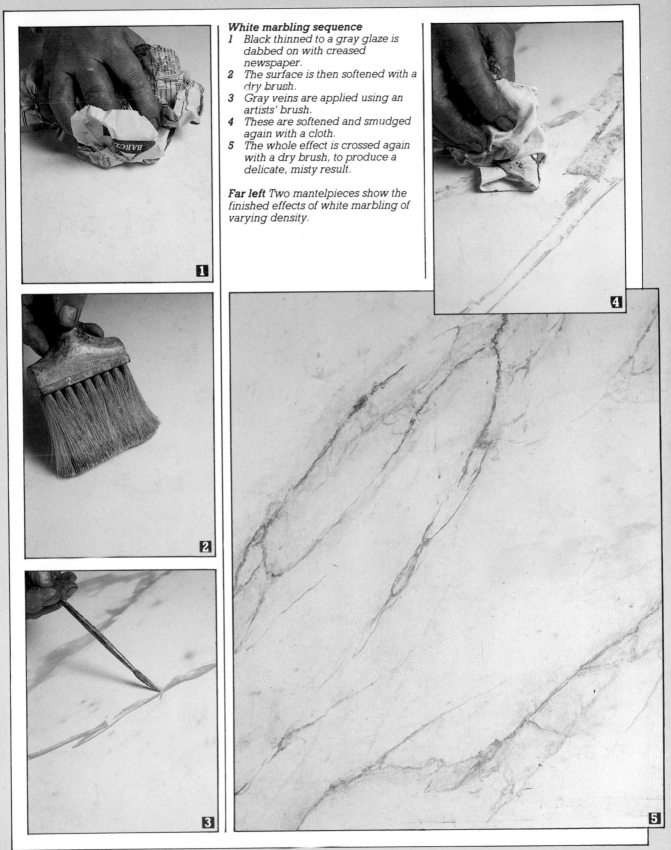

White marbling sequence
1 Black thinned to a gray glaze is dabbed on with creased newspaper.
2 The surface is then softened with a dry brush.
3 Gray veins are applied using an artists' brush.
4 These are softened and smudged again with a cloth.
5 The whole effect is crossed again with a dry brush, to produce a delicate, misty result.

Far left *Two mantelpieces show the finished effects of white marbling of varying density.*

Above These marble pillars are restrained and their color is well balanced with the rest of the decor; more opulent coloring would make them overpowering.

Left: Marbling a pillar
1 The base is a creamy white ground color.
2 This is softened with an ocher glaze, ruffled by using newspaper to apply it.
3 Veining is applied with a feather or brush.
4 The whole effect is softened with a further glaze, smoothed off with a dry brush.

you may end up with an effect less like marble than an infra-red map of the earth.

■ **Tools and materials** For all types of marble effects, you will need the same basic equipment: flat-oil, egg-shell or undercoat and a transparent oil glaze, in appropriate colors (*see below*); artists' oil in a corresponding color; oil crayons; a small marine sponge and a feather or a soft 2in or 3in (5cm or 7.5cm) paint brush , for "softening"; saucers, clean rags, screw-top jars and mineral spirits.

WHITE SICILIAN MARBLE

■ **Application** There are three basic techniques for this marble. For each of these, the base color is dead white — applied in flat-oil, undercoat or egg-shell — rubbed down with glass-paper to eliminate any brush strokes or unwanted marks. The surface should be perfectly level and smooth, as marble doesn't bulge. The other materials for this effect are artists' oil color in Chinese or flake white, raw umber, black and yellow ocher, and oil crayons in black and dark gray.
Method I First, blend a little of the white oil color with mineral spirits

until it is a near-transparent mixture. Then, separately, dissolving the oil color in mineral spirits and adding an equal amount of glaze, mix a small amount of each of two glaze shades, a yellowish gray and a greenish gray. For the yellowish gray, mix white plus raw umber plus a little yellow ocher and black; for the second, white and raw umber and a little black. Then, taking a rag, rub the thin, white oil mixture all over the surface of the prepared undercoat base color, so you get a milky, off-white effect like cloud.

Next, draw in the veins with the wax crayons, using the gray one for the softer, inner veins and the black for the harder, outer ones. Let the vein pattern meander diagonally over the surface. The pattern is rather like a bolt of black lightning striking across a white sky. Remember that marble veins spread like a tree's branches, to left and right of a central node on the stem; so, on a wall, imagine the sparse crown of a tree upside down and then another spread of bare, twiggy branches rising from the floor until they all meet and join up. Or, if you like, picture it as a series of raggedly diagonal rivers, all running across a map; but remember, the branching tributaries of these rivers *always* join up again eventually; they never appear from nowhere nor do they just peter out to nothing. Be as free and sweeping as you like, but take care not to overdo it; always err on the side of too few lines rather than creating the vast web of some manic spider. There should be big, wide areas, across which these lines meander with wide spaces between. The only crowded or busy areas should be where one set of veins crosses another. Put the pale gray lines on first and then follow them like ragged tracks with the darker shade.

When you've finished veining, sponge the large white areas with the gray-green and yellow-gray glazes; do this sparingly and don't entirely cover the white ground. Then, with a broad, dry paint brush, strike the whole surface diagonally one way, and then the other. This will make the crayon lines and glazes blend softly, and the whole effect miraculously seem to resemble marble.

the brush gently as you draw it across the surface toward you, varying the pressure to vary the width of the line. Soften these lines a little with the sponge and transfer some of the darker color on the sponge to other areas adjoining the lines, to achieve a delicate bruising effect. Then go over the whole surface again in one diagonal direction with a dry brush.

Method III Replace or augment either the wax crayons of the first method or the artists' brush of the second with a long feather. Dip the edge of the feather in water and then in mineral spirits, and then comb it to separate the fronds. To create the veins, brush some oil color onto the fronds and draw the feather across the surface. When you turn it side-on to the wall, the feather will leave a sharp, single line; tilt it, and the frond lines will branch in all directions; turn it back on edge again and the lines will coalesce sharply into a single vein again. This is most effective if the soft veining colors are used in conjunction with wax crayon. You can create a very soft and subtle effect by crossing the veins afterwards with a dry brush.

ROSE MARBLE

■ Application

Method I This marble can have a base color of soft, sandy gold — beige with yellow added and a spot of red — with a pale pink glaze brushed over it. This glaze can be applied so that there are corridors or canals where the base color is left to show through. These unglazed, linear patches may be quite wide — up to 6in (15cm) — so that they form wide, sandy veins across the rose-colored areas. Then, mix a glaze of 4:1 parts blue to red, with a touch of white and a trace of burnt umber to give a gray-purple. Sponge or brush this loosely over the pink-glazed areas. After this, dab the surface with a crumpled newspaper, to give a soft, creasing effect. Veins of thin white can then be traced onto the surface with a feather or an artists' brush, crossing over the purple/pink areas as well. The whole surface is finally softened again with a dry brush.

Method II A matt coat of deep pink or gray-pink — magenta, white and a touch of black or gray — should be laid on over a light gray base coat, in the same way that the white ground is laid over the base coat in Sicilian marble. This should be over-glazed

Above *Rose marble and pink-veined white: two effects achieved by reversing the same colors, rather like a photographic negative. Note the broad fluidity of the veins on the right and the delicacy of those on the left.*

Right *One of the most sophisticated of all marbled finishes. This effect is achieved by highlighting the lower sides of veins and brushing the upper sides upward with a feather and dry brush, then over-brushing a pale amber or gold glaze.*

Method II This technique avoids the laying of the white oil over the base coat as in Method I, and uses just the white base coat instead. With a mixture of 1:2 parts of paint to mineral spirits, mix a paint glaze of raw umber and black, which should have just enough color to show against the white ground. Sponge this over the whole surface. Then, using two different mixtures of raw umber and black, to give a light shade and a dark shade, mix the two tones for the veins. Taking a very slender artists' brush, hold it straight out as though it were an extension of your hand and, keeping your palm upward, rotate

Top *The exotic element in marble combines admirably with objects of fantasy.*

Above *Tone co-ordinates these two colored marbles, the darker adding visual weight to the stairway, and stressing the sweep of its form.*

Left *A large area of stylized marbling, which works well because of its subtle coloring and rhythmic simplicity.*

with a thin white paint glaze, followed by an ocher glaze laid in rough, feathery diagonals. These should be softened with a dry brush and then, either with a feather or artists' brush, a deeper blue and red glaze should be applied to evoke veins, roughly following the ragged, softened, ocher veins. To finish off, the whole surface may be softened with a dry brush.

BLACK SERPENTINE MARBLE

The first method given for simulating serpentine marble is simpler to execute than the Sicilian effect, except that it needs to be done horizontally because of the flow of the paint. The second is perhaps rather less easy, but allows a vertical approach. In either case, black serpentine is a very beautiful and dramatic marble and should be simulated sparingly. In small amounts it can be superb, *en masse* it tends to be somewhat oppressive.

■ **Application**
Method I For purposes of simulation, at least, serpentine has a black ground, mottled with a dusty emerald and streaked with random, thread-like veins of white. Lay the ground with black, oil-based paint, then mix the top coat of green paint glaze — emerald with a touch of raw umber and black to dirty it or take the edge off it. Mix this in a ratio of 1:2 parts mineral spirits to paint, so that the glaze has a reasonable body. When you've applied the glaze, take a stiff brush and splash mineral spirits at random all over it by flicking the brush. Where the mineral spirits fall irregular apertures will open in the glaze to reveal the black ground, a process known as cissing, an effect that can also be obtained by flicking water over oil-based paint. Next, squeeze a marine sponge out in thin white paint glaze and dab it on the areas of black ground revealed by the cissing, so that an occasional mottle of white results. For the veins, it is highly effective to coat lengths of cotton thread or thicker twine in white oil color and lay them on the surface to produce a random flow and an occasionally crossing fretwork of fine, diagonal lines.
Method II This involves using oil- and water-based paint together, and enables you to achieve this effect on a vertical surface. A black, oil-based ground should be applied in the usual manner, and over this a black oil-based coat, thinned in a ratio of

1:2 parts mineral spirits to paint. This should be laid on so that there are still patches exposing the dry base coat and also thin, linear areas of veins joined to these patches. While this is still wet, unthinned latex in white or very light gray can be taken up very thickly with a marine sponge and dabbed lightly on the surface where the wet, black glaze coat has not been applied. You will find that latex will flow into the wet edge of the oil. Using either a feather or a thin artists' brush, held in the manner described for Sicilian marble, roll the latex along the linear vein spaces left in the oil coat, fairly liberally, teasing it into the edges of the oil coat. You will have islands of hazy, swirling mottle where the sponge has been, linked by hazy veins of white. Add very fine lines by laying cotton strands, liberally coated with latex, against the wet oil surface; hold them at both ends and slap them against the surface. The natural action of the water-based paint floating on the oil surface often means the surface doesn't need brushing afterwards, but a dry brush can be used to soften areas where the effect is in any way short of the desired finish. Over this, when dry, should go a thin, emerald glaze. When this has dried, the surface should be given a clear coat of matt or semi-gloss varnish to prevent a different texture occurring and catching the light between the two types of paint. Brushes used for this process should be washed out in mineral spirits and then warm, soapy water.

RED MARBLE

Rose marble has a soft, glowing quality to it, whereas red marble has a deep, sumptuous appearance, the difference perhaps between silk and velvet. Like black serpentine, the red marble pattern is exotic and best used sparingly. Its ground is an orangey magenta, rather like a punch made of red wine and pineapple juice. The grain or vein of this marble evokes a ragged white net or the diamond-shaped mesh of a wire fence that is encrusted with ice and torn in places.

■ **Application**
Method I Over a dark gray or brick-red base coat, brush or sponge a ground color of 4:1 parts deep pink to yellow ocher with a touch of blue, thinned by about a third with mineral spirits. Mix an off-white glaze — flake white and yellow ocher — and, using a feather, simulate a wide-spaced

Above *A green marbled, varnished mantelpiece adjoins a ragged amber glaze chimney-piece and a yellow Sicilian marble.*

Varnishing Marble

All painted marble should be varnished to protect it, and polished stone generally has a satin finish. Apply two coats of clear gloss varnish and one of satin to walls and woodwork, and four coats of gloss plus one of satin to floors. A tinted varnish applied to white marble gives a very attractive, deep, translucent tone and results in an aged appearance, often found on mantelpieces and marble-topped furniture.

Right: Sequence for green marble
1 *Broad ruffles of green glaze are applied over black with a feather.*
2 *The surface is cissed with mineral spirits, sprayed from a bristle brush, to open up the pattern.*
3 *A mixed green and white paint glaze is feathered cross-wise over the preceding green veins.*
4 *Fine white veins finish the effect.*

version of the mesh pattern. Soften this pattern by crossing the edges of it lightly with a dry brush, leaving the central strokes of the veins fairly crisp. When this has dried, go over the whole with a thin, light blue glaze. Finally, in the center of the junctions where the veins of the grid cross, use an artists' brush or half a potato to add flecks of deeper magenta or magenta and blue.

Method II This is a similar technique to that of using oil- and water-based paint to simulate black serpentine. After applying a brick-red base coat, apply a second of the same colour with a trace of blue and a touch of yellow ocher. While this is still wet, float an off-white, ragged mesh pattern of latex over it in the manner used for black serpentine. When this has dried, the veins may be strengthened with a feather.

Method III This is an exact reversal of the preceding technique. Over the base coat of brick-red, apply the ragged vein mesh with off-white, oil-based paint, using an artists' brush or large feather. Then fill in the areas between the veins with latex. Where you brush or sponge the latex up against the oil-based veins, a soft swirling and blending will occur, and fine wisps of the oil-based paint will be carried like threads into the latex. This is perhaps easier to apply than Method II, as oil-based paint is more dominating than water-based, and in the second method there can be a danger of the water-based veins being "drowned" in the oil surface and disintegrating. This third method is also rather quicker, as the quick-drying latex means that at least two-thirds of the area is dry and can easily be touched up while the remainder is blending; while in Method II, the whole surface is liable to be wet for a considerable time before you can strengthen any areas that are not satisfactory.

GREEN OR TERRA VERDE MARBLE

This is probably one of the most attractive of marble patterns, as it has the panache of black serpentine but is less stark. Its salient features are a moss-green ground with shifting, deep blue clouds across it, veins of ocher and gold and then serpentine veins of white crossing the deep blue-green. Occasionally it may have a gold flecking, which can be added by a very fine spattering.

■ **Application** Over a base color of flat-oil, undercoat or egg-shell in a dark- or mid-green, apply a paint glaze of 3:1 parts artists' oil paint to mineral spirits, to give a blue-green clouded effect. This oil color should be 3 parts ultramarine blue, 1 part emerald and 1 part burnt umber. Brush or sponge this glaze liberally all over the base color, leaving sizeable areas uncovered by it. Then, using an artists' brush or a feather, add the yellow veins: they should consist of an oil glaze with 3 parts yellow ocher, 1 part red and 1 part white, and be allowed to meander as they do in Sicilian marble. Cross the veins with a dry brush to soften their lines and, if necessary, strengthen the inner parts of the veins afterwards. Then, using lengths of cotton thread soaked in white oil paint, lay sinuous, white veins across the blue-green areas between the yellow veins. The yellow veins should be much broader and more powerful than these thin, white lines; all are diagonal, but the white should appear like little shoals of eels passing through wide, green spaces in a gold mesh net.

If you want to create the effect of gold flecking, take a dry 3in or 4in

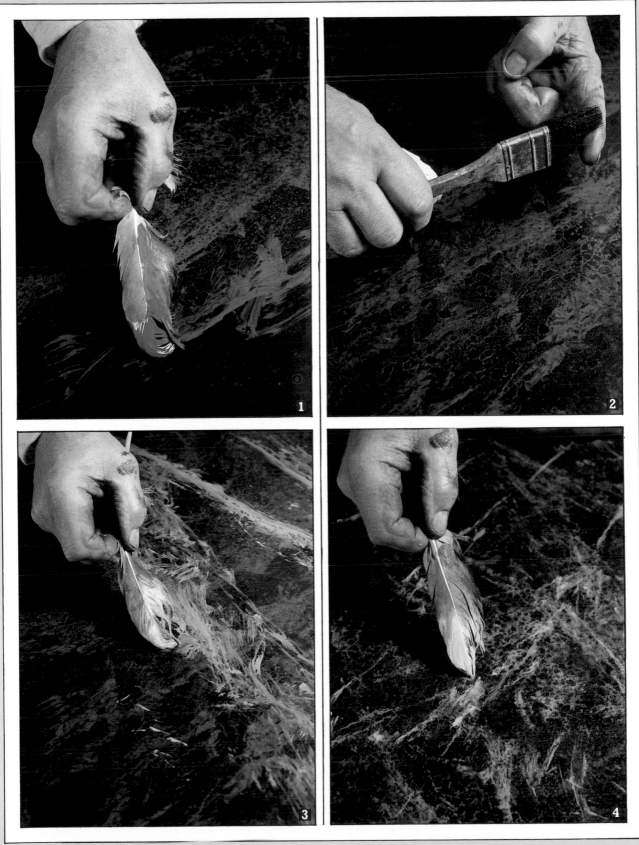

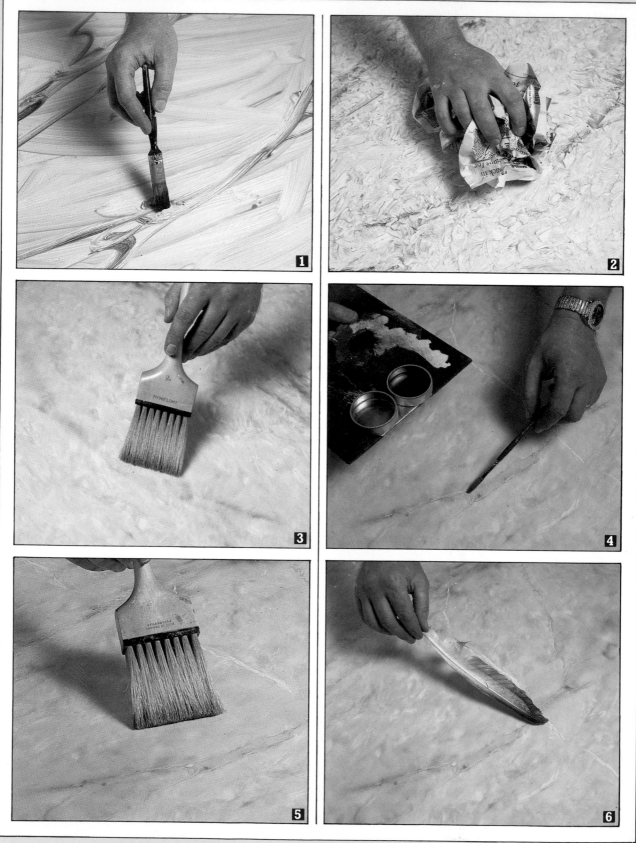

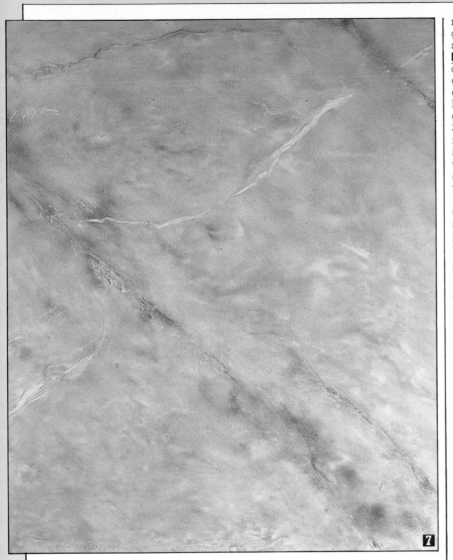

mid-blue, and over this a series of glazes is applied, more evenly spread than for most marbles.

■ **Application** Mix a thin oil glaze of ultramarine, 1:2 parts paint to glaze, and brush it all over the base color. Then, using a feather or artists' brush, lay swirling veins into this glaze, in an off-white paint glaze — 3:1 parts paint to mineral spirits. This resembles Sicilian marble but is more fluid, like cigarette smoke. Over these and the blue areas, apply another thin, blue glaze so that the whole effect is pale and deep blue. Next should go a slightly thicker colored glaze of magenta, leaving some patches uncovered to give a purple-blue, bruising effect; then, using cotton lines coated with white paint, add thin white lines along these bruised areas. Soften the whole with a dry brush, giving a thin, overall glaze of emerald as a finishing touch.

OTHER TECHNIQUES

The preceding marbles are an example and guide. To begin to enumerate all the possible colors and variations of this stone would be to recite the color wheel in the form of veins and grounds and base coats. There is simply no limit to the colors you can choose and the only limits to the technique are set by the materials you use. With an oil-based paint glaze, mixed 1:1 with mineral spirits, you can virtually marble on any ground in any color. You can mix three or four glazes of different colors and spread them over the ground and blend them, dabbing the whole surface with a marine sponge or creased paper. By adding transparent oil glazes of lighter and darker hues, you can create a feeling of depth in the finish that is peculiar to marble as a polished rock and gives it that element of clouds and tinted veils that is the prime secret of its light-retaining quality. Spattering any color in showers of fine flecks can be effective, provided that it isn't overdone, as most marbles do not include too many of these isolated crystals. However, spattering mineral spirits or wood alcohol in the same way — even on a vertical surface — is almost always highly effective, as it opens the glaze for a cissing effect. It also works very well on blue and rose marbles. A cut cauliflower floret is a superb simulator of the large, branching crystals of milky white found in the pink marbles, and

Left and above: Sequence for yellow marble

1 *An ocher glaze is brushed loosely over a cream ground and rough veins of dark brown added with a smaller brush.*
2 *The surface is ruffled and blended with crumpled paper.*
3 *The whole is softened with a dry brush.*
4 *Blue and sepia glaze veins are added with an artists' brush.*
5 *These are softened with a feather or broad brush.*
6 *Fine white veins, unsoftened, are applied with a feather tip to cross the existing pattern.*
7 *The fine white veins appear to float on successive layers of color.*

(7.5cm or 10cm) brush and coat the ends of the bristles with gold enamel paint. You may need to thin this paint slightly with mineral spirits, about 1:4 parts mineral spirits to paint but no more. Hold the brush 6-12in (15-30cm) from the surface and run a steel-toothed comb across the bristles in a single saw-stroke, to give a soft spray. The flecked areas occur in the blue-green patches among the thin, white veins. It is somewhat easier to use this method than a canned spray, which tends to give too dense a delivery to the surface for this marble.

BLUE MARBLE

This is an effect like deep blue, clouded glass. The base color should be a flat-oil, undercoat or egg-shell in

some reason, they are reluctant to flow, dab more mineral spirits on with cotton batting. When they have dried sufficiently, veining over them can be achieved by any of the methods already mentioned, particularly with cotton twine lines, although this has to be executed flat.

■ **Paper rocking** One of the quickest, most effective methods of simulating marble, and very subtle and gentle in appearance, is moving newspaper or another thin, easily crinkled paper over the surface glaze. A thin paint glaze of just off-white or a very delicate blue-gray or pale ocher-gold can be evenly spread over a base of matt white; the paper, which must be creased or pleated beforehand, is then laid against this. Where the creases come against the glaze they will raise or absorb it, leaving fine, vein-like lines. If the paper is then pushed or stroked with a stiff brush against the surface, it will shift the glaze surface, blending different colors together. This rocking of the paper will give a softer, more subtle transition and sense of movement than any other equally quick method. If the glaze is a soft monotone, such as pale gray-blue or ocher, the result will be highly realistic when the paper is removed. If the glaze is varicolored, because the fine vein lines will cross different shaded areas without relating to them, the result will be less realistic but very pleasing, evoking the general visual atmosphere of marble rather than its actual appearance.

MARBLING FLOORS

Wall-paints, artists' oil colors and specialized floor-paints are all suitable for floor marbling. In the case of the first two, the prime difference lies in the number of coats of protective varnish necessary — about five. Other than that, any of the techniques mentioned are suitable, with one or two logical considerations to bear in mind. For instance, if you are going to apply a very fine, delicate marble effect, well protected by varnish, such a delicate finish is going to look very odd on floorboards where the planks are more than $1/16$ in (1.5mm) apart. You are never going to come across marble arranged in long, board-like slabs side by side, not to mention veins that mysteriously jump over the cracks to the next block. Either you must fill the gaps between the boards, or you must apply a

Above Sponged and cissed gray marble is constructed with a blue-gray paint glaze laid over white, with veining in the same white as the base, and a blue glaze applied over the mottling.

Top right A matt white Sicilian marble is highly effective on a large, solid furnishing.

Bottom right and far right Marbling before and after a coat of tinted varnish, which protects and mellows.

loaded with off-white paint or used dry to remove glaze it has few equals as a tool.

■ **Floating color** So far we have discussed only those effects using dry grounds. Floating means marbling on a wet ground. The base color should be dry, of course, but then a coat of flatting oil — a mixture of 1:6 parts boiled linseed oil to mineral spirits — should be brushed over it. Onto that should be brushed a really thin solution of the ground color, diluted with mineral spirits until positively watery. Using two or three other colors to the same level of dilution, and ¼-½ in (6-12mm) brushes, dab these colors on the wet ground. They will flow together with ease — in fact, so easily that you may have to dab them with cotton batting squeezed out in mineral spirits. If, for

Above *A fine example of a wooden floor, painted in the form of marble tiles and varnished. Note how the border of the floor pattern is the same width as the baseboard is high, and its presence solves the problem of getting awkward quarter-tile widths against walls and corners.*

much bolder, freer pattern to the marble, so that it is obviously a marble *effect* — in the same way that a book cover may be marbled — rather than an attempt at direct simulation of the real thing. Similarly, you should be careful of the coloring you choose. A bathroom or kitchen floor given a marble pattern with a white ground and blood-red veins will conjure up all sorts of gory images. On hardboard or chipboard floors, or on floorboards with the spaces well filled and sanded, the methods used for walls and other woodwork can be followed. On a

floor, it is wise to use three coats of gloss or semi-gloss varnish topped by two coats of matt; this gives a sheen and a sense of depth to the color and evokes the natural properties of marble surfaces. The darker marble finishes, such as red or serpentine, look well on floors because of their visual weight.

■ **Floor-paints** Specialized floor-paint is an effective marbling medium. Even its limited range of coloring is no drawback, as the colors can be inter-mixed and the muted tones are more of a help than a hindrance. There are two ways of mixing. Firstly, the base color can be applied at full strength and the veins, thinned 1:3 mineral spirits to paint, applied with a brush or two or three feathers tied together (to cope with the thicker quality of the paint). The base color is then applied again in the spaces between the veins and the two wet edges blended and softened together with a feather or dry brush. The alternative technique is to apply the base color and the veins simultaneously, using one brush for each color, and then blending the two together with a feather or dried brush. A dark gray ground with deep brick-red veins works well, or a deep red ground with gray or white veins, a white ground with gray and/or green veins, a red ground with gray or blue-green veins, black with white and green, or blue-gray with white and/or deep red veins. Coating cotton twine lines with paint and then laying them briefly on the base color is a particularly effective method of veining on floors.

TORTOISESHELL

As a decorative medium, real and simulated tortoiseshell originated in the Far East, with the first costly examples reaching Europe in the seventeenth century. The great majority of these were small and exquisitely prepared: panels mounted in ivory or ebony, small lacquered boxes or toilet utensils. Soon, tortoiseshell's distinctive coloring, markings and opulent appearance made it a prominent feature in the craze for Things Oriental that subsequently swept Europe. Cabinet-makers and decorators were not slow to reproduce its appearance, on objects ranging from lacquered furniture to cornices and ceilings — where it appeared in the form of inset oval and circular panels. For the most part, these areas were kept small; this was partly an aesthetic decision and partly due to the medium required — varnish. Varnish is essential for tortoiseshelling, but it tends to dry quickly, making it almost obligatory to restrict the finish to small areas. The limited size of the original examples meant that they could be sumptuously exotic and rich, which made them very popular. European craftsmen of the period soon realized this and, although they

rarely achieved the finesse of their oriental counterparts, whose varnishes were long perfected for this use, they often showed greater variety and innovation, while taking care not to apply excessively large areas of this exotic shell patterning.

Bearing all this in mind may help you to avoid making the worst aesthetic mistake possible with tortoiseshell: applying too much of it over large areas as if it were marble, or over areas of intricately curved and convoluted moldings or heavy furniture, where real tortoiseshell could never have been used. Such

Above *Painted marble, simulating the type of jointing and patterning typical of a floor or table top. The dark, crystaline patterning forms a good frame for the open, fluid-veined yellow, and the painted joints have a thin, pale fillet to imply a filler between the slabs.*

an error of judgment results in transforming a beautiful, often jewel-like mottle into a disastrously vulgar absurdity.

The natural coloring of tortoiseshell varies from golden honey tones through the tawny auburns to an almost fire-ember red. Even so, like marbling, the term tortoiseshell has come to describe a particular type of paint finish, a patterning and use of paint and varnish in the manner of tortoiseshell but without necessarily being an exact copy of the real thing. Tortoiseshell can, for example, mean imposing markings of deep brown or red-chestnut over a deep blue or emerald. In this way it may even resemble the patterns found on butterfly wings.

The scale of the patterning you use should be most influenced by the size of the area you intend to paint, and that area should be either a flat or a gently curved surface, not one with a high relief or moldings. Because the technique involves brushing patches of oil paint into wet varnish it is very difficult for one person to tortoiseshell a large area — which is just as well, as it prevents invidiously large-scale execution with excessive results. If tortoise-shell is chosen for a large wall, it is wise to divide the area into panels — no larger than 3ft (90cm) — and to execute the panels alternately so that each has its own identity, rather than creating one big spread, which will look awful. Two people can do the job more speedily than one but remember that (as with shading and stippling), each person has a distinctly different touch, so either one person should always apply the varnish and the other brush the paint into it, or they should each work on their own separate panel section. If you decide to divide a wall into panels, divide it first with chalk lines, then put masking tape down the edges of each area. Paint the panels alternately, leaving every second panel blank; when the first set is dry, paint the others, removing the masking tape when you've completed both sets. Paint the areas between the panels afterwards, using masking tape along the edges.

If you intend to work on a door, it is often useful to remove it from its hinges and lay it down flat. Doors are a prime example of where and where not, and how and how not to apply tortoiseshell. If the door is panelled, it is very unwise — and very unsuccessful — to tortoiseshell the entire surface. It will simply look silly and ugly because the panels and raised areas together will bear no relation to the patterns of the tortoiseshell. However, if you paint only the inset panels — and they are small, plentiful and regularly spaced — it can look very handsome. A flush door can look superb. It is also possible to simulate the kind of surrounds that were occasionally used to frame tortoiseshell: metal, ivory and ebony. Ivory can be simulated by an Indian yellow and flake white oil paint covered with two coats of varnish — a matt coat over a gloss coat — both tinted to a pale amber. Ebony can be evoked with burnt umber and lampblack, given two coats of varnish — matt over gloss — both tinted with a trace of red.

Whichever way you may choose to evoke tortoiseshell, always remember that it is an opulent, exotic effect. It works well on flat details in bathrooms, intimate halls, dressing rooms or bedrooms, on furniture — especially the tops of small, decorative side-tables — and ornamental objects. Like marble, it should not be applied to a surface that could never have been made of tortoiseshell in the first place, or to which a panel of credible size could not have been fitted. Giant tortoises may be substantial creatures but no elephant-sized ones have been located to date.

GOLDEN TORTOISESHELL

This is probably the lightest of the realistic tortoiseshell effects, essentially a blond rather than auburn finish. It gives a sensation of less weight and can therefore be executed over a slightly larger area.

■ **Materials** The ground or base coat should be an egg-shell or low-luster, oil-based paint in a light, warm, sandy yellow. Tinted varnish is essential: either dark oak thinned 1:1 with mineral spirits, or light oak. On plaster, varnish is better unthinned, so it is preferable to use undiluted light oak; on wood, thinned varnish is more effective. Two dark brown artists' oil colors, raw and burnt umber; two ordinary paint brushes, one for wet varnish, the other to be used dry, sizes 2-5in (5-10cm); two artists' brushes about ½in (1.2cm) in width; mineral spirits, clean rags and cotton batting are also necessary.

■ **Application** It is a very good idea to have a look at a sample of real or

1

2

Sequence for golden tortoiseshell
1 *A tinted varnish is brushed over a cream ground and irregular, Indian red patches twirled into it simultaneously.*
2 *These are crossed and blended.*
3 *Indian red and burnt sienna patches are teased into the varnish with an artists' brush.*
4 *These are blended with a swift, crossing stroke.*
5 *The whole is softened by criss-crossing with a wide, dry brush.*

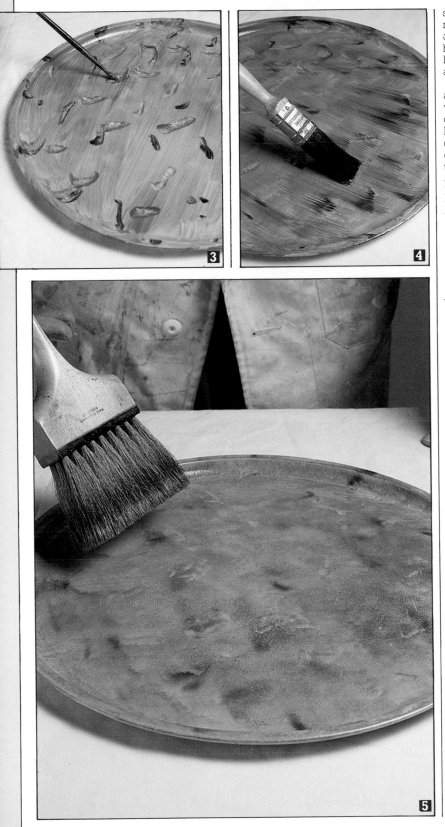

simulated tortoiseshell, in either a museum or publication, before attempting to reproduce it. It does have a very distinctive appearance, but is subtle and various for all that, and by no means repetitive.

The base coat should be applied and allowed to dry. Then, before varnish is applied, the tinting oil colors should be prepared with a little mineral spirits. The movement of tortoiseshell, like marble, is diagonal and slightly radiating, so that bands of color diverge. To achieve this effect, use one of the two broader brushes, and apply the varnish quickly and liberally over the base coat, moving diagonally from the top right (unless you are left-handed, in which case start at the top left) down over the whole area of the panel. Then, using the same brush, work or tease the varnish into a series of diagonal bands of zig-zag direction and irregular width. At this point, take either a screwed-up rag or a piece of cotton batting soaked in mineral spirits and fray the edges of the varnish bands so that they will be receptive to paint blending. Now, before the varnish sets, work the lighter of the two prepared oil colors into the gaps between the bands of varnish with one of the two artists' brushes. For this, use either a flat, zig-zag motion — as if you were shading with a pencil or wax crayon — or roll the brush-tip over the surface of the varnish. Then insert the darker brown into the center of the light with the other small brush.

Once you have applied these colors, soften their edges into the varnish by stroking the whole surface very gently with the other large brush, which you should have kept clean and dry for this purpose. Go in the same direction as that of the dark patterns; then cross off vertically, and finish off by going in the original diagonal direction again. You can repeat this procedure as often as you like until you get a soft finish that satisfies you, but always remember that the final diagonal should be in the direction in which you originally applied the varnish.

This is the most basic tortoise-shell technique, but you can add as many accents and flourishes as you like. Flecks or freckles often occur in tortoiseshell along the lines of the main marks, but set at a slight angle to them, rather as pine needles radiate along the general line of a twig — only, of course, much more sparsely. Fill an eye-dropper with mineral spirits or wood alcohol and let spots fall sparsely on the varnish. This cissing will open the varnish and into the resulting holes you can insert a small spot of the darker oil color as soon as the cissing has dried. Augment the cissing with small, dark spots of oil color and then drop a spot of the solvent into these to create rings, or eyes. When these eyes have dried, you can add a tiny spot in the center of them, if you like this effect. Alternatively, you can fray the edges of the bands or spots of oil color in the same manner, and leave the cissing spots unfilled if you choose.

AUBURN TORTOISESHELL

This is a rich, red-chestnut tortoiseshell, reminiscent of the tawny wings of the tortoiseshell butterfly.

with the lampblack, 2:1 paint to varnish or, if you prefer it paler, 1:1. Use this mix to insert the first, broader, diagonal paint strokes in the usual manner. Then brush a mix of lampblack and burnt umber into these, concentrating the deeper tones at the center. Blend the paint and varnish with the dry brush as normal, and then, in patches along the line of the main paint marks, ciss the varnish to your taste; make sure that the cissing occurs just alongside the paint marks but not over them. Fill the cissing with black or burnt umber.

A darker, red version of this variant uses a solid, brick-red ground, dark oak varnish and lampblack oil color; its method of application is the same, mixing varnish with oil color in a ratio of 2:1 for the initial dark strokes.

AMBER TORTOISESHELL

This is a warm, golden-tinged effect, deeper than the honey-blond tones and less red than auburn. It has the same visual weight as mid-oak panelling.

■ **Materials** The base color should be yellow ocher and the varnish dark oak. The two oil colors are raw umber and lampblack.

■ **Application** If you intend to use a large, dark, zig-zag pattern with this tortoiseshell, you may want to thin the oil paint in varnish for the initial paint shading after the application of the main varnish coat. If you do this, the undiluted raw umber should form the center of the zig-zag with a trace of black. If the dark paint areas are smaller, the raw umber need not be diluted but simply brushed into the varnish in the conventional manner, with the lampblack used in the center. Cissing in the center of dark paint spots — to make rings — works well on this version, as gentler contrast prevents the softness of the rings from being overshadowed by other elements.

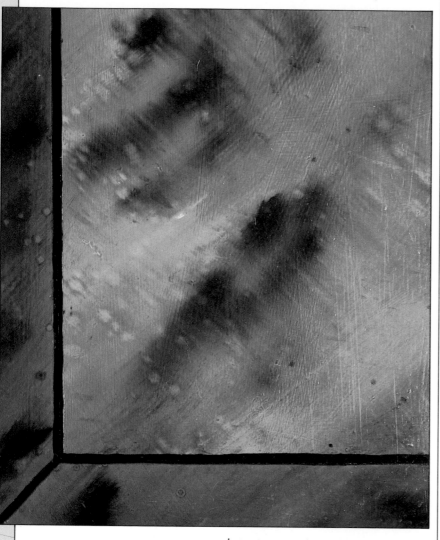

Above Painted golden tortoiseshell panel with painted auburn tortoiseshell border. The lighter area has patches of blue brushed into it as well as burnt umber, and cissing spots, splashed on after the finish has been crossed with a dry brush.

■ **Materials** A base of brick-red mixed 1:1 with yellow ocher, a dark oak varnish and two artists' oil colors — lampblack and burnt umber — are necessary to apply the finish, in addition to those brushes and tools previously mentioned.

■ **Application** Over the base coat, apply a dark oak varnish, using the same type of stroke as you would for golden tortoiseshell. Before applying the oil color, mix some of the varnish

FINISHING OFF

Absolutely never use matt varnish to give a top protective coat to tortoiseshell; it kills the light-reflective quality in the paint/varnish. Strictly speaking, it isn't really necessary to put on a top varnish coat at all, as the technique is essentially a paint/varnish one anyway, but real tortoiseshell has a sheen like silk. A semigloss polyurethane varnish is effective, but the most accurate and

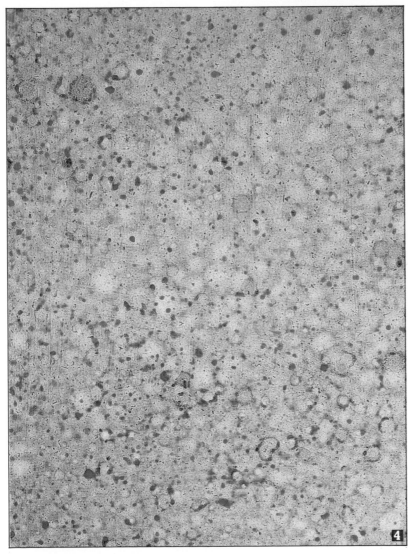

Sequence for gray porphyry

1. A gray, glazed ground is brushed with criss-crossing strokes to give it a marked texture.
2. It is then criss-crossed diagonally with a soft-bristled stippler.
3. Amber and terracotta flecks are spattered over the surface by striking the brush haft on a wooden batten, and then cissing with mineral spirits.
4. The final finish closely resembles granite.

perhaps the most pleasing way to finish tortoiseshell on woodwork is to sand down the polyurethane top coat with a very fine wet-and-dry paper, used with a solution of water and mild soap-flakes. Then polish this with a solution of rotten-stone and warm linseed oil, mixed to a paste.

PORPHYRY

The name 'porphyry' comes from the Greek word for purple, the reddish purple associated with the Caesars. The term was and still is used to describe the red-purple variety of an igneous rock, which can be polished like glass and was frequently used for figurative sculpture. But this hard, granitic rock with its crystalline drifts occurs in many color variations: dark green flecked with gold and black; violet flecked with gold, black and iron-gray; red-brown flecked with light brown and black, or with red-purple, black and pale pink; brown marble-veined with almost translucent white and flecked with pink, red and green; or gray-green with flecks of white and black.

Porphyry as a paint technique is really a specialized variant of spattering. It is easier than marbling, once you are acquainted with the technique, and is applicable to wall areas of most sizes, especially those of bathrooms and small hallways, floors — provided the patterning is not too pale — and the tops of some furnishings, especially side-tables; it can be used contrastingly on baseboards and mantels. It can be far lighter in appearance than tortoiseshell and, as with marbling, there is no real constraint on the colors you may use. However, even when used with free color combinations, the technique still evokes a stone-like quality and not just a pleasing, but random, paint finish like spattering.

PINK OR CINNAMON PORPHYRY

This is a deep cinnamon-colored rock, flecked with pink, red and green, and sometimes with smoky veins of milky white.

■ **Materials** The ground color should be flat-oil paint or undercoat tinted with artists' oil color in beige or mid-brown; use a dark red and trace of black with the beige, an orange and black with the brown, and blend 1:6 parts oil color to paint. Also necessary are flake white and opaque oxide of chromium (mossy

gray-green) oil color, mineral spirits, ordinary large paint brushes for the base color and smaller (2in and 4in (5cm and 10cm) ones for spattering. Optional extras include oil glaze and an artists' diffuser; this is a very simple piece of equipment, consisting of two small, hollow tubes set at right angles to each other, which you can get inexpensively from any artists' supplier. A small artists' brush or large feather is also required.

■ **Application** When you have tinted and applied the base color, allow it to dry thoroughly. Meanwhile, mix six variations of the spattering colors. All should be diluted with mineral spirits in a ratio of 1:3 paint to solvent. From the base color, make two pale coffee browns by adding flake white, one slightly darker and "dirtied" with a spot of lampblack; two greens, one by adding flake white to opaque oxide of chromium, the other flake white and a trace of red to the opaque oxide; two pinks, one by mixing flake white and red, the other flake white, red and a trace of opaque oxide. Also make a 1:3 paint to solvent dilution of flake white.

If you decide to include a white vein pattern, take either an artists' brush or a feather and, in the manner of marble veining, apply a loose, swirling, slender sinew across the surface — as if you were imitating a wisp of smoke. This veining should meander and it can split, but it should always ultimately go to the edges of the surface, never just stopping. If you do not think the effect is soft enough, use a dry brush to cross the veining as you would in marbling, then commence spattering. Whether or not you apply the veins, begin spattering with the deeper of the mixed oil colors: the deeper green followed by the deeper pink, then the paler green and pink, and finally the two off-whites. You can, if you choose, apply a glaze between some of these coats, to increase the depth of the surface effect, but this isn't essential. Remember, though, that porphyry is a crystaline rock, and the spatters which evoke the mineral crystals don't have to be as regular as, say, the spattering of a non-representational pattern that relies on a more even, stylized finish. But don't overdo it; you rarely see porphyry with a mottle larger than ¾in (1.8cm), the great majority of sizes ranging between wheat grains and fuzzy rice seeds up to garden peas. Often there are fine, mottled

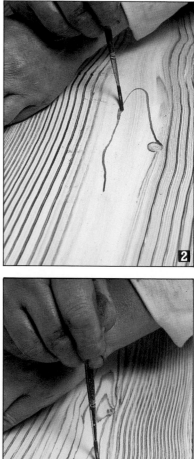

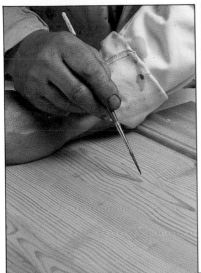

Sequence for wood graining

1   A paint glaze is brushed on freely, and then either a wide or narrow comb or two or three fingertips wrapped in a soft, clean cloth remove the paint in grain-like, wavering parallels.

2   Finer grains are drawn in with an artists' brush or sign writers' liner over an area wiped clear with a cloth.

3   Knots are touched in with a fingertip and then strengthened with an artists' brush.

4   The fine grains are thickened into long ovals around knot and heartwood sections.

5   The final finish of this method is an unrepetitive, almost photographic realism.

clouds like the Milky Way, and these are very easily created with a diffuser. You put one end of the tool in a pot of thin paint and blow gently through one of the two remaining apertures. This draws paint up to the junction of the two tubes; another puff dispatches a fine rain of paint from the remaining outlet onto the surface. It is quite straightforward, so long as you remember not to suck. For larger spatters, strike the haft of a loaded brush against a flat piece of wood; for the finer crystal spots, load a stiff-bristled brush at the tips with the spattering paint, then draw the teeth of a stiff comb or a knife tip in one sawing stroke across the bristles, creating a fine spray.

If you are working on a flat surface, the finish can easily be cissed by spattering mineral spirits over it. This opens up the paint and, if you have applied a glaze, it will produce an appearance of bursting and expanding crystals where the solvent strikes the spattered paint spots. This is also possible on a vertical surface, but keep a rag handy to soak up any dribbles quickly.

This basic method of application is the same for any porphyry, regardless of the colors used. Remember that you always lay dark spattering colors first, with the lighter shades laid on top to give the surface a greater effect of depth.

Porphyry has a variety of finishes. It can be semi-matt, like roughly worked granite, or it can have a glass-like, high gloss finish — that enamel-like polish seen on much finished statuary. On walls, it looks best matt or with a soft sheen but on smaller areas such as baseboards, panels and furnishings, it can have a more reflective finish. In all cases, porphyry should receive two coats of clear varnish; otherwise, when you clean it, polishing will damage its fragile spattered surface. A satin finish probably gives it the best degree of shine.

WOOD GRAINING

Like so many other decorative techniques, the craft of simulating wood grain dates from Ancient Egypt, a land traditionally short of wood; the technique was thus developed to imply the presence of a costly material, hard to obtain. Centuries later, in Europe and elsewhere, similar techniques arose to imitate rare exotic woods. These techniques have been developed to

a high level so that the combined understanding of paint and of the properties of wood have produced simulations indistinguishable from the real article. Many of these involve the manipulation of tinted varnishes, similar to the technique for tortoiseshell; others require over-glazing oil paint, then applying a coat of varnish. Most of these almost photographic simulations are outside the scope of the amateur, but there is really no need to attempt them. As with marbling, the chief asset of graining in most interiors is its general effect, the sense of texture,

warmth and that organic harmony that is peculiar to the grain of wood.

Woods are so various and the contrasts of their grains so great that most graining techniques set out to take the general principles of wood's structure and treat them in an informal, but recognizably organized, way. Graining as decoration is seldom, therefore, a study of a particular wood — be it pine, beech, rosewood or mahogany. These four are so different that they are an ideal example of the futility of attempting a generalized technique, aimed at reflecting all four.

Above *A fine example of delicately tinted wood graining on floor and window paneling. The variation of the grain is considerable, but all the sections have the same visual weight and the result is a highly sophisticated, quiet unity.*

If you intend to simulate wood grain, it is a very good idea to look at real wood for a while and study the way it has grown or "flowed" before you begin work. Notice how the heartwood shows long, ragged ovals with the grain widely spread and softly defined, forming broad areas, not sharp waves. See how grain usually runs roughly parallel, while rippling gently and rhythmically; how it grows darker as it approaches a knot and swirls around it, and then pales as it passes beyond. Notice that at times the grain draws apart, is broader or narrower, and how even at its straightest is never really

straight. Like rippling water, it is infinitely various, but it does have a specific, unmistakable pattern; it is never anarchic. These are the general features of softwoods and the majority of hard, though in hardwoods the grain is usually much finer and closer.

In the majority of cases, it is useful to use a series of tones of the same color when graining. Establish the general tone of the wood, say, light oak — a pale gray-gold — and then, choosing a base color slightly lighter than the intended finished color effect, add the grain in slightly darker tones. Even so, it is quite easy to produce contrasting grains of a different color, or to over-glaze wholly different colors on top of those used for the graining itself. Contrasting grains can be as gentle and subtle as silver-gray oak — that is, a dark gray ground with a trace of ocher, with the grain in a pale gray or off-white. With careful choice, a pleasing grained effect can be produced in colors that have no natural connection with wood itself, such as amber grain over emerald, or emerald grain over deep gold.

The essential structure of the grain can be stylized and conducted with a considerable degree of fantasy; but if you exercise this licence, guide it with knowledge. Take a rubbing of a wood grain with a wax crayon or charcoal on paper to give you a graphic view of its structure and make you familiar with its more prominent markings. Also, coast a dry brush over the grain of wood to give yourself a feel of its flow and way of growth. Close your eyes and allow your touch to dictate to you or, still with closed eyes, run your fingertips over the surface. Then emulate this motion with your hand away from the surface. Having said all this, there is no need to become too studious about graining as the ultimate aim is a decorative effect, not a photo-realist painting.

BEER AND VINEGAR GRAINING

Either oil or water-color paints can be used for graining, and beer or vinegar mixed with water is probably the cheapest version of the water-colors. The ground for this technique needs to have a flat finish, so flat-oil paint or undercoat is best, sanded down with wet-and-dry abrasive paper and soapy water. This sanding should be carefully and thoroughly done on egg-shell paint to ensure good adhesion and to flatten the sheen. The chief problem with using water-color on an oil ground is that globules of water tend to form on the surface rather than spreading in an even film. For this reason, it's advisable to run over an oil ground with whiting on a damp sponge and then dust off the loose whiting powder when dry. Rubbing the ground with soap solution will also help to eliminate grease spots that will otherwise cause color to ciss.

Stale beer — preferably brown ale because of its high sugar content — mixed at a 1:2 beer to water ratio, is capable of giving a variety of shades from deep amber to pale gold when applied with a brush. Malt or red wine vinegar and water at the same 1:2 ratio gives a slightly redder tone (don't use onion vinegar — it is too pale); it's a good idea to add a little sugar to this solution to help it stick to the surface — about 2tsp:1pt (10gm:600ml). Either mixture can be tinted with powder pigments by mixing the pigment to a smooth paste with a little of the beer/water or vinegar/water mixture first and then gradually stirring them into the main body of liquid.

These beer or vinegar stain/ paints have a translucency that can give a remarkably sophisticated finish if they are used with care. In fact, the sap that produces wood grain is not dissimilar in consistency. The softness of the colors also means that they can be strengthened with additional brush strokes without building up a sticky layering, which is what tends to occur when thickening other types of paint. They can be blurred softly along the edges with a damp sponge, and it is even possible to evoke the ragged ovals of heartwood by laying on a long oval area of water with a sponge and then coasting a brush loaded with beer or vinegar stain down each edge. The inside edge of the stain stroke should haze and blend, fading in toward the center of the oval.
The other grains can be applied with an artists' brush, from 1/16in (0.15cm) upward, a ½in (1.2cm) being about the best general width for washing in between the grains, and a ¼in (0.6cm) the most suitable for the grains themselves.

■ **Application** There are two main methods for beer or vinegar graining. The first has much to do with the basic techniques of all water-color painting, and the second with dragging — a method also used for oil-graining (*see below*). The water-color approach involves drawing in

Sequence for simulating bird's eye maple

1. A generous paint glaze is laid over the ground and a series of rhythmic crinkles achieved by using the end of a soft rubber comb, pressing right down, then releasing sharply upward.
2. This effect is softened with a badger softener.
3. The "bird's eyes" are blotted in the wet glaze with a fingertip.
4. The eyes are joined with a soft pencil in the manner of grains.
5. The result is a highly stylized version of this maple.

Above *This effect was achieved by paint glazes applied as on page 107 but using a grayer tone. Then the cross-grains were scraped out of the wet glaze with a clean, round-edged tool, such as a pencil wrapped in cloth.*

Three Decorative Woods

Rosewood A rich auburn paint glaze should be brushed evenly over a mid-brown ground. With a feather or a ¼in (6mm) artists' brush, grains of deep burnt umber should be laid on in sweeping parallels that swerve jaggedly sideways for a few inches, at roughly 18in (45cm) intervals. Soften with a broad, dry brush.

Walnut A soft, deep, honey-brown paint glaze should be applied evenly. Then with a feather or ⅓in (8mm) artists' brush, veins of deep coffee (2:1 burnt umber to raw sienna) should be added to look like a shoal of eels wriggling side by side, each between a few inches and 2ft (60cm) long. Soften the overall effect with a dry brush until the pattern is foggy.

Oak Evenly apply a gray-brown paint glaze of 2:1 raw umber to burnt umber. Extend grains of burnt umber outward from a knot in concentric ripples on one side of the knot, and on the other extend them as wavering, parallel lines. For these veins use a ¼in or ⅛in (6mm or 3mm) artists' brush.

Left *One would not readily imagine that this very stylish side cupboard could be finished just by mottling its tinted glaze with a folded cloth and then softening the effect with a dry brush.*

the grain lines with a brush, in the same manner as you would use a pencil; a sign-writers' brush, known as a pencil writer, is highly effective for this. Load the brush to about a third of the way along the hairs, but not so much that it drips. Holding it as you would the shaft of a pen, rest the tip on the surface and, in a continuous movement, draw the brush downward; the good news is that you don't have to worry about wobbling, but try not to let the stroke appear "tight" — full of little kinks. This will happen if you hold the brush too tightly and go too slowly. Practice on some paper first. To soften one side of a piece of grain, either take a larger brush, dampened but not dripping, and tease the edge of the grain stroke; or lay a stroke of water down the surface before you make the grain stroke and let the grain line follow along the edge of the wet area. Keep a sponge handy at all times. Knots can be simulated by laying a circular patch of color on the surface and pressing a dry, notched cork immediately into the wet area, or a torn, screwed knot of blotting paper, which is often more effective.

There are two great assets to this method, besides being inexpensive. Firstly, it dries very quickly, in about 15-20 minutes, which means you can touch it up without worrying about smudging the parts that you have already done; and secondly, that it can be washed off immediately after application if you haven't got what you want, provided that you've sealed the surface first. Some people consider the quick drying a drawback, as you may not achieve the desired effect before it dries; but this scarcely matters since you can still wash it off when dry. In any case, you can slow the drying time by adding a few drops of glycerine. Probably the prime limitation of this medium is its delicacy; like so much water-color work, it looks best fairly small. To undertake a very large area can make the effect appear too fragile for the space, so it is best to stick to door panels and small, inset areas.

It is absolutely essential to protect water-color of this type with at least two coats of clear, semi-gloss varnish, otherwise it can easily wash off or wear away.

GLAZE GRAINING

This is graining in oil, using a transparent glaze over an egg-shell ground. The ground should be hard, smooth and non-absorbent; if you use a flat-oil ground, you should give it a protective coat of clear shellac or satin-finish varnish before graining. The basic technique involves brushing glaze over a ground color and then giving the glaze a texture with a graining comb or a dry dragging or graining brush.

■ **Materials** The correct tools are essential for graining. These can be expensive and so, even if you don't wish to splash out for the real thing you absolutely must get the nearest approximations.

You need a medium-sized decorators' brush for initially applying the glaze color. Then a graining brush; you can easily make one yourself from a stiff, thin-bristled brush by chopping the bristles off square about 1in (2.5cm) from the stock with a craft knife and a hammer, and then cutting out clumps of bristle from each side of the brush, alternating the clumps and leaving a slight space between them. You will also need a graining comb; you can make this from notched plastic, linoleum or stiff card, or use a cheap metal comb if you don't mind the grain being very parallel — you can also use this comb for separating brush bristles. A notched cork is useful for knots, a wide, soft-bristled paint brush for mottling, and a fine-pointed artists' brush for touching in individual lines of grain. Also necessary are clean rags, mineral spirits and, preferably, a sponge.

Professional decorators use a veritable army of brushes for graining, the vast majority of which you can manage without, if you use those suggested above; but there are four of greater importance. It is a good idea to get two of them if you can, as they will make the work far easier and give it a more sophisticated finish. One is a flogger , which is used to beat the surface and "fizzle" the grain; that is, to give it kinks. A mottler is another, which mottles and gives the surface highlights. A badger softener is used to blend and soften effects of all types; this is a very old type of brush that comes in a great variety of sizes and was originally made of badger hair. Fortunately, the protection of badgers in some countries has meant that these excellent tools are now made of other materials that are very nearly as good; although this does make the term "badger softener" rather mysterious to many people. Of

these three brushes, the badger is undoubtedly the most useful and versatile and if you have to choose one of the three, choose that one. It can basically be used for all three effects; flogging by using it side on, mottling by using it vertically or in a pencil-shading motion, and softening by stroking; you can even drag with it if you keep it dry. The fourth decorators' brush is a fine sable writer — a long, fine-pointed brush used by sign-writers and sometimes known as a pencil because of its fine point. Sable is pricy, but there are nylon substitutes available. These purport to be as good but they aren't — they have an unpleasant habit of fanning out and going hard. The sable writer is really very useful; it is about the same price as other artists' brushes of the same size and is definitely a recommended investment.

■ **Application** Bear in mind that you are basically creating the essence of wood in this effect, rather than making an exact representation of a real, specific wood. Choose the elements you like most about any particular type, such as the light, flowing grain of pine or the close,

flecked grain of beech, and stress those. Always remember to keep a balance between the absolutely regular pattern — which will look unnatural and is the great, lifeless weakness of wood-patterned wallpapers — and the completely random design, which will be just as unrealistic and also unpleasantly chaotic.

Method I If you apply a glaze tinted darker than the base color, you can comb it, but waver the comb as you go so that the teeth leave fine, undulating bands of the base color while the grain stands out in a darker tone against them. This may be quite sufficient for your taste, but if you want to go further you can touch up these combed lines with a dilute solution of 1:2 paint to mineral spirits. If you seek a more subtle grain, use a graining or dragging brush. This is applied much in the manner of the dragging technique of broken color, in this case to remove wet glaze. The result is a fine, fluffy grain if the drag brush is well dried. A denser, heavier effect comes from a cut-down brush but if you use a comb to separate out the bristles evenly, from time to time, an excessive weight of

Above *Wood graining used in conjunction with stenciling. The arching patterns of the stenciling balance the regular angularity of the wood, and the wood gives a visual anchor and stability to the stencil.*

Right *This room shows widespread use of painted wood grain. Frequently, the real wood of doors and panels may be in poor condition, stained, unsightly, and often of different types of ill-matched wood. Over-painting it, and then giving it a painted grain, enables you to give any impression of age and wear that you wish, and to highlight variety where you want it.*

stroke is easily avoidable. If you beat the graining glaze with a flogger or dusting brush substitute, you will get fine openings in the glaze like pores. Hold the brush horizontally to the surface and beat lightly and rapidly as if beating a fragile rug. If you strike along the line of the grain and then across, you will get a closely textured, fibrous quality. Alternatively, you can hold the brush perpendicular to the surface like a stippler, or push it backward or forward over the grain to achieve a softening of the grain markings. Thirdly, you can apply a glaze of clear shellac over either a painted grain or a textured glaze, and then finely spatter color over the shellac surface, softening it immediately afterward by whisking it with the tip of a soft, dry brush — such as a badger — in the same direction as the grain.

These are all textural methods. You may find any of them effective but none are indispensable. You can texture a glaze, let it dry and then paint grain on with a dragging brush, using the bristles to put paint on the glaze, rather than to remove it from the base coat. This is part of Method II.

Method II Using a brush to put on colored glaze rather than to remove it allows greater versatility in the effect. The movement is more fluid and the grain can be varied more easily in width and intensity. In one movement — if possible — draw the brush down the surface, undulating it as you go; this is much easier than formal dragging, as you don't have to worry about keeping the stroke straight. Where you feel a knot should be, turn the brush at a curving tilt so that the spread of the bristles becomes narrower, and then curve it back again as you straighten out after completing the bulge. That will give you a tightening of the grain around the knot, which then visibly opens up again as the grain flows on beyond it. Knots are best applied with a notched cork or a torn, screwed-up piece of blotting paper: coat the end of the cork or paper with paint and press it flat against the surface, then withdraw it. A second time, on another spot, twist the cork or paper slightly; another time, only half-coat the end or rotate and/or run it slightly downward about 1 in (2.5cm) in a short, sharp action to create a long knot. Another method is to core the bristles of a stenciling brush and then rotate the brush on the surface, but the effect is less

convincing than that produced by a notched cork or blotting paper.

Wood frequently possesses a mottled, silky quality, and the most straightforward way to evoke this is to load a cloth or chamois leather with mineral spirits (or water, if you are using a water-based paint) and dab or roll the surface with it, usually between widely spaced grains so that it frays their edges. Alternatively, dab mineral spirits on the bristles of a dry brush and draw them lightly across the grain, rocking the brush from side to side.

Method III This has more in common with marbling than with dragging and combing, and its finish is ultimately far more sophisticated than the preceding methods, with perhaps the exception of very finely executed beer graining. The materials are a large painters' brush for the application of the base coat, three hues of artists' oil color, depending on the tone of the wood — for example, burnt umber, lampblack and Indian red for rosewood — mineral spirits, linseed oil, rags and three artists' brushes, preferably a ½in (1.2cm) bristle flat, a sable from ⅛in (0.3cm) to ¼in (0.6cm), a sable writer and a badger or feather.

Above *A painted red mahogany pillar adds weight and strength to this corner, dividing the figurative mural from the marbled wall. The wooden pillar links these effects logically and discreetly.*

Left: Sequence for mahogany
1 *A red-brown paint is applied and stirred in sweeping swirls, using a soft dragger.*
2 *Darker, burnt umber grain swirls follow the pattern, applied either with a dragger or with a fine brush, using a scribbling or shading stroke.*
3 *A short dragger gives a silky ripple to the grain.*
4 *The whole surface is softened with a badger softener.*

The base coat should be flat-oil or undercoat, tinted with artists' oils to the required tone. The graining itself is executed with three sizes of artists' brushes or sable writers; the oil color is diluted 2:1 or 3:1 paint to mineral spirits or, if the surface is such that drying time is not a problem, then 1:1 or 2:1 oil color to linseed oil. The surface can then be brushed overall with a 2:1 mixture of linseed oil to mineral spirits, or with the faster-drying preparation of flatting oil — 1:6 parts boiled linseed oil to mineral spirits. The filaments of the grain can then be brushed or floated onto this surface, being softened into it where desired with a feather or badger softener. Or, conversely, the grain can be laid on with the small brushes, and the spaces between them softened or blended where desired by brushing their edges with a 2:1 linseed oil to mineral spirits, or flatting oil, mixture. These methods produce a smoother surface and a great sense of depth and translucency, such as can be seen in polished woods such as rosewood, mahogany, cherry and satinwood. Of course, such techniques take longer to dry than the preceding ones, but their finish is vastly superior. However, it is better to practice first, and they demand a degree of patience. If such a method is used on a door, it is a good idea to unhinge the door and re-hang it when dry.

Once this finish has been allowed to dry, it should be given two or three coats of low-luster clear varnish, or the varnish can be rubbed with a solution of rotten-stone and warm linseed oil to give it a soft polish.

BAMBOO

Bamboo, like tortoiseshell, appeared in Europe in the seventeenth century. It reached the zenith of its popularity in the eighteenth century when, being so costly, it was often imitated by craftsmen in turned wood with plaster of Paris joints. With the extension of imperial power and a subsequent lowering of cost in the early nineteenth century came the craze for bamboo furniture — this time in genuine bamboo. Both real and simulated bamboo are still popular today, and bamboo is imitated on both wood and metal. Real bamboo, dating from its revival in popularity during the nineteenth century, is often found in thrift shops in need of renovation, which

Above and top *The basic color of natural bamboo is yellow ocher. On a molding, furnishing or picture frame, you have the freedom to stress the darkness of the joints as much as necessary; if they look better pale, imitate male bamboo; if you wish the effect to be heavier and smokier, choose female.*

Right *Superb early 19th-century, cast iron, stylized bamboo banisters at the Royal Pavilion in Brighton, England. The sense of height is increased by the long, female spines on the single columns. Note the disk patterns that loosen the formality.*

means cleaning and frequently also painting.

Real bamboo that needs decorating should be washed first with wood alcohol to remove any grease; then, if you intend to paint it overall, give it two or three coats of thinned, flat-oil paint or undercoat (the same consistency as for walls), sanding it between coats with wet-and-dry abrasive paper and soapy water. If you are dealing with imitation bamboo in turned wood, a wash with wood alcohol will do it no harm and will eradicate grease; then seal it with a coat of clear, matt varnish or shellac.

It is relatively straightforward to repair and paint over simulated bamboo that has damaged plaster of Paris joints. Plaster of Paris is very similar to surgical plaster — indeed, sculptors use the two in exactly the same way — and it can be bought in packets. Although it isn't very cheap in small quantities, to repair such a small area it's not worth buying a 1cwt (50kg) sack, especially as it goes off quickly after it's been opened. Plaster of Paris or surgical plaster are both mixed by gradually sprinkling the powder onto the surface of water (to repair a small joint, you can mix enough in a saucer) and adding it steadily in small sprinklings until the powder ceases to sink but forms a thin film on top. Never stir this plaster when mixing it; it causes air bubbles. When the powder rests on the surface the plaster will "go off" and as soon as it acquires the consistency of heavy cream, apply it to the joint with your finger and let it set. It will go hot and steam, then go cold. Let it harden for about eight hours, and then you can carve the joint with a craft knife. Give it a rub with wet-and-dry abrasive paper, and then a coat of shellac. Then you can paint on it.

Male and female bamboo are distinguished by their joints. Female bamboo has a ring on each side of the central joint knot, shaped in profile like a pair of prim lips. On each side, above and below the lip joint, it has two spines about ¾in (1.8cm) apart, that curve up and meet and then extend away as a single spine of variable length into the section above the joint. At times the tip of this single spine overlaps with the tip of the one coming down from the joint above, but they are usually asymmetrical, and it is better if you paint them so. Between these, above the lips, are two little oval eyes,

darker than the lips and lighter than the spines. Male bamboo has two thin, sharp, semi-translucent rings, one on each side of the knot, which is itself darker than the rest of the cane.

■ **Materials** The best paints to use in simulating or renovating bamboo are the strong, quick-drying sign-writers' colors or artists' acrylics. To paint female bamboo, thin them to a creamy consistency, smooth-flowing but opaque, using mineral spirits for the sign-writers' paint and water for the acrylics. For the sharp, semi-translucent male lines, use oil color thinned with a little mineral spirits and mixed with varnish, and then further thinned with solvent. You will need three tones to paint the joint of male bamboo. Mix the thinnest of three tones first, 1:1 paint to solvent; then the two darker, with more paint and varnish, the darkest being about a 4:1 mixture of varnish to solvent. You will need two ordinary paint brushes for male bamboo, one about ¾in (1.8cm), one ½in (1.2cm), and an artists' sable of about ⅛in (0.3cm).

■ **Application** The overall color of real bamboo is a light yellow ocher; the knots darken to a delicate amber, while the lips are umber and the spines and little eyes a smoky umber with a trace of lampblack or deep green. In male bamboo, the knot rings are translucent, extending about ½in (1.2cm) on each side of the center. The rings can be applied with an ordinary ¾in (1.8cm) brush; when the first ring is dry, a second, darker ½in (1.2cm) band should be superimposed on it. Then, using an artists' brush, you can apply a narrow, dark band in the center, so that the overall result is like the smoky joint you see on lightly charred wood, or a gazelle's knee. The same colors apply to female bamboo, with the eyes as the darkest part. The artists' brush is pressed into the lip-band, following the contour in the same way that lipstick follows the contours of a mouth. Painted bamboo should be varnished in satin or gloss to your own taste, although satin most closely resembles bamboo's natural sheen.

In the Far East, bamboo was and is frequently painted in brilliant colors — most commonly vermilion and gold, brilliant blues or jade green — and picked out in contrasting tones that center on its natural ring and knot formations. Whether you choose natural or brilliant colors, the cleaning order and preparation remain the same.

6

FINISHING TOUCHES

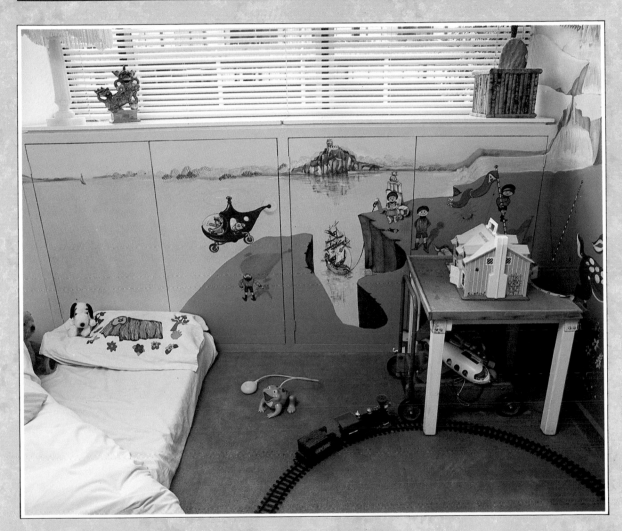

These are decorations that complete the effect of a room, taking their
cue from existing features. Lining and picking out frame and highlight
all forms of broken color, marbling and wood-graining. *Trompe
l'oeil* creates painted illusions, implying dream-like 3D; stenciling
and glass painting offer infinite pattern and variety.

should therefore be fairly thin. If you are using a light lining color over a darker ground, test it first to ensure that it is fluid enough yet still has adequate covering power. It needs to be smooth or loose so that you don't have to keep stopping because the brush is tacky, as that will give a jerky line, a little like shriveled bamboo.

■ **Materials** Artists' oil colors or universal stainers, mineral spirits and varnish, two artists' sable brushes, a flat-ended ⅛ – ½in (3-12mm) for broader lines and a rounded 1/10in (2.5mm) for thinner strokes are a good combination of materials. Professional decorators use flat-ended hog-hair fitches; beginners would find the shorter ones easier to use because they have less give , but longer ones are rather better as they hold more paint and therefore give a smoother, more continuous stroke.

■ **Application** Dissolve the artists' oil color or universal stainer in a small amount of mineral spirits and mix the solution in a little varnish to give it greater viscosity and to keep the bristles or hairs of the brush close together and therefore running more smoothly. Goldsize is an alternative to varnish and is quicker drying but rather stickier. If the mixture is too thick, thin it with mineral spirits. Linseed oil is really better, as it enhances the color, but it is very slow drying. It is a good idea to strain the color, too, because otherwise you'll find particles of undissolved pigment floating in it, which can cause a sudden darkening of stroke.

Don't be intimidated by the idea of lining because the lines have to be relatively straight; it doesn't necessarily demand a steady hand. Lining is a simple handicraft, not a technical drawing skill, and makes none of the demands of graphic or fine art. Part of its charm stems from its small irregularities and some minor waving of the lines makes it look more "alive"; it's easy enough to practice on board or lining paper before you start, and most decorators never paint a line freehand anyway.

For short, straight lines, all you need to do is turn the beveled side of a ruler to the wall; the bevel will hold the edge clear of the surface and avoid smudging. For longer lines, glue corks to one surface of a long straight-edge to hold it away from the wall, and maintain a straight line-stroke by resting the middle finger of your brush hand on the

Top Lining corresponding to the darkest tone used in the area it surrounds, lending the whole a sense of definition without heaviness.

Above The cream lining enclosing this pale area is of the same tone as the lightest part of the ragged darker area below, providing visual unity.

Left Lining used dramatically over a largely flat ceiling and wall area conjures up illusory steps and terraces, heights and distances. A finely balanced adventure for the eye that stimulates but never bewilders.

LINING

Lining is the straightforward term for a finishing technique that gives cohesion and form to wall and woodwork areas; these areas may have been decorated in any of the broken color techniques described previously, marbling, rolling, spattering or dragging. At its most simple — and, frequently, most effective — it consists of painting lines or bands of color around or across a given area to sharpen or stress its basic shape. It includes picking out , or enhancing the three-dimensional appearance of raised features and, at its most sophisticated, overlaps into the area of *trompe l'oeil* — simulated three-dimensional effects on a flat, two-dimensional surface.

When lining is being practiced as a simple decorative band of color and is not intended to deceive the eye, the choice of color used is entirely personal, but as a general principle it should relate either to the main color of the wall — say, a tone or two darker — or to some dominant color in the general decor. If it is being used as a containing line — on marbling or tortoiseshell, for example — it should pick up the darkest color in the pattern. Lining color needs to flow freely, and

straight-edge as you draw the brush down or across the wall.

Before you apply the painted lines, it's useful to establish their position with chalk lines, checking that the lines are in the right quantity and thickness. Don't leave too much chalk on the surface, or it will pick up on the lining brush; make sure you remove the surplus just before you start to paint. If you are going to join two lines with a curve, draw the curve you want on a piece of card and then puncture this curve line through the card with very small holes. Place the card against the surface, then run the brush over the card and lift the card off; you will have a curving row of very fine, painted dots. Then all you have to do is join the dots up, carefully. (This is how the outlines of figures were painted onto the Sistine Chapel ceiling.) Rest your little finger against the surface to steady your hand when painting the curve over the dots.

It is quite helpful to apply a coat of clear varnish to the surface first, as this will enable you to wipe off any slips or large errors with a rag moistened in mineral spirits. The only real difficulty with lining comes from leaving a slither or blot unremoved on a semi-porous surface. If you do allow this to happen, you should sand it off gently, making sure that you don't go through to the surfaces beneath, as retouching the ground will almost always show. However, if you do have to retouch it, just feather the patch gently into the surrounding area, using your finger, and blend it further, when dry, with fine abrasive paper.

In applying the color itself, try not to overload the brush or you may find it leaves blots. Press the brush gently against the side of the paint container to release any excess and then try to apply the paint in a long, single stroke. Start as far up as possible, keeping at a sufficient distance from the wall to give your arm freedom of movement, and try to remain as relaxed as possible. Rest your hand on the straight-edge and, keeping your eye on your hand and the straight-edge — not the brush — go on down or across the surface in one movement. Avoid stopping during the stroke, as this will cause more of a variation in the texture. Even if there is a wobble, finish the stroke and then go back and wipe it off at the offending area. Stopping and starting causes heavier and lighter areas of paint, and it's easier

Left and top Two examples of recessed planes and moldings. The foremost have been crisply lined, those at the back softened to recede.

Above Moldings starkly lined to balance sharp mesh.

Left page The elegance of classic form is poised as much on careful tonal balance as on proportion.

to touch in a longer area in one smooth stroke than lots of blobby little ones. The pressure you exert and the amount of paint you have on the brush will dictate the width of the stroke, rather than the brush itself.

Once you have become used to this method, you can use deliberately faded (thinned) paint lines, or vary the weight of the lines by darkening some sections with a smooth second application. Another way of fading lines is to sand them over slightly to soften them and then over-glaze them when dry.

TROMPE L'OEIL

Trompe l'oeil is the technique of creating visual illusion. The earliest celebration of it is Pliny's well-known story from the fifth century BC of the good-natured contest between the painters Zeuxis and Parrhasios. Zeuxis painted a picture of grapes so well that birds flew up to peck them; then, seeing the curtain covering his colleague's picture, he asked for it to be drawn back. Parrhasios replied that the curtain was the picture. Zeuxis immediately ceded the prize to him, for he had succeeded in deceiving not only the birds, but also Zeuxis, an artist himself. Of course, the story is apocryphal and to judge visual art merely on its capacity to simulate three-dimensional space is naive; indeed, many cultures, particularly those of China and Japan, have long held such things to be aesthetically childish ; but as far as decorative painting is concerned, *trompe l'oeil* is a skill of the highest order. In fact, it would be wrong to infer that the amateur can produce such effects with ease; many are far beyond the skill of all but a small number of artists. Nonetheless, it is surprising how much can be achieved by the lay person through a variation of the technique of lining, especially when it is augmented by some of the techniques previously described in this book.

The raised moldings of doors, baseboards and inset panels can all be simulated on a flat surface. Recesses can be conjured up in the same manner, and an illusion of texture and space created where otherwise there might be only a bland, flat surface. These effects involve very simple manipulations of light and shade, using just two tones of color closely related to the base color of the surface.

Above *Lighting is most important to trompe l'oeil. Painting a shining lamp directly beside a light source rarely works.*

Far right *Here, the high, three-dimensional effect of the plant and vase, book and mask seem to float on a more two-dimensional table; the result makes the wall appear illusory.*

Lining on Furniture

Lining is suitable for many types of furniture, apart from those which are highly polished and ornate. On stripped wood, which needs priming, the lining pattern should be applied in primer, if necessary using masking tape, and the finishing coat added over it. On sponged furniture, the outlining of drawers and table-tops sharpens form, especially if a contrasting color of the same tonal value as the darkest or lightest of the sponged colors is used. By applying the lining pattern in tape to stripped wood before sponging, and removing the tape afterward, you can obtain a marquetry effect. Lining without tape gives a freer appearance and fine lines are best guided with chalk or pencil. Turned wood can be lined around the raised rings, but discreetly, unless you're after the barber's pole effect. This technique works well in conjunction with stenciling, expecially when bordering the loose folk patterns often applied to stripped wood furniture.

Above A non-existent kitchen closet. Less photographic than some effects, the plates and loaves look dreamlike on real shelves — except that the "real" shelves aren't there, either.

Left The standpoint of the observer is often crucial to trompe l'oeil. The dramatic perspective through succeeding arches drawn on a wall is made more credible by the size of the building. The painted horizon is a trifle high, because if the observer could look down on it, it would not work.

■ **Materials** Most paint is suitable for this process. You can use artists' oils, diluted with mineral spirits; artists' acrylics, diluted with water and, if necessary, mixed with their own gel retarder; flat-oil, egg-shell or undercoat tinted with artists' oils or universal stainers; or latex tinted with gouache, acrylics, stainers or powder colors. Generally speaking, you need the same brushes as for basic lining — a ½in (1.2cm) flat-ended one being the most useful general size — a straight-edge, chalk or a sharp pencil, and masking tape.

■ **Application** The basic technique is the same, however complex the ultimate effect may be. The

elaboration is really dependent on one's own imagination, and you may wish to copy a series of panel moldings or carved surfaces. Even if you decide to copy a Rococo scroll by tracing the design and squaring it up, the lines you paint will be dictated by the effects of light and shadow around the outline of the three-dimensional form. The safest approach is to err on the side of simplicity.

When you have decided on the design, how big it is and where you want it, mark out its position with chalk or pencil and, if it consists mainly of straight edges, use a straight-edge to draw it. If you are evoking geometric raised moldings

Above Trompe l'oeil *as dream-like fantasy. The stylized, poetic blue sky found in mid-1960s gallery painting and the red sash window are dream imagery, as are the birds; they're frozen in flight and so can't fool the onlooker into believing them real, but the effect has seductive, picture-book charm.*

Right *Nothing here looks really three-dimensional. The quirk of this interior is to make almost everything — including the furniture — look two-dimensional, which gives an altered sense of reality. Even the light source and time of day seem uncertain.*

Far right *Another example of the difficulties of lighting* trompe l'oeil. *The paintwork on the left is far more successful than that on the right. Had the angles of the walls been made into painted sea-side awning poles and the ceiling corners into a beach-house eave and awning, the illusion would be more credible. As it is, it's only partly successful.*

surrounding panels, each panel will require a rectangle of parallel lines about ½in (1.2cm) apart, with mitered corners. Tint your paint to give two shades, one just a little bit darker than the ground color, the other darker again by the same proportion. Now decide where your light source is positioned — to the left or right. A simple track molding should be shaded as follows: if the light is striking from the left, you shade the right side of the right-hand track lightly, and the left side of the left-hand track darkly. This gives the effect of a highlight striking on the left side because of the greater contrast there. If you wish to keep the tracks straight, use masking tape to make the miter joint sharp. Set the

tape at 45° to the rectangle and paint right up to it on one side with the appropriate tone for the area, then remove the tape. When that tone is dry, reapply the tape at the same angle but so that you can paint up to the 45° join from the other side, with the other tone. The tape should shield the tone you've just done. This technique of creating light and shade can be taken to any degree of elaboration by careful repetition. It can be applied around or over rag-rolled, stippled, spattered and broken color areas; over or around marbling, and around tortoiseshell and dragged areas, but not usually over them.

If you want to give the impression of a deep recess, say a rectangular

one, lay in the shape of the implied rectangular molding and then, on the inside of the inner line of one of the sides, extend the darker of the two shades inward for about a third of the way. Then draw a vertical to terminate it from top to bottom of the rectangle and stop the shade at this line. Next, at the lower corner of the side you've shaded from, extend a line upward diagonally toward the center of the panel and, where it crosses the vertical you have just drawn, draw a horizontal line across to the opposite inner side of the panel. Then fill in this area with a slightly lighter shade of the ground color.

Within the area of these painted panels you can add anything you

wish. They are particularly useful for inserting mirrors, an area of differing texture to the rest of the decor — such as porphyry amid marbling — or even paintings. Stencils also fit well in them, and so does graining.

PICKING OUT

Picking out means heightening features that actually exist in three dimensions. On public buildings, this technique takes the form of painting moldings and other raised surfaces in colors that contrast with or offset the main coloring of the interior or exterior. In domestic interiors, this is rarely advisable. Such a treatment can become overpowering and make a lounge look like the central offices of a county court. Used carefully in coloring closely related to the main base colors of a room, however, the picking out of features can greatly enhance a domestic interior. There are two methods for this process, sometimes referred to as positive and negative , the first using paint and the second, tinted glaze.

■ **Application**
Method I When using paint, the first consideration must be the shades used throughout the rest of the room. Remember that moldings and raised details are an intrinsic element of the whole — they are not the chief objects in the room, nor the whole point of its decoration. It is safest to use a tone of the same color family as the walls or a major furnishing, or the color of dominant woodwork. Your choice should be further simplified when you bear in mind that a raised feature painted in a darker color than its surroundings is going to turn into a silhouette; whereas if it is lighter, the light striking it will cast shadows that outline its features, thus enhancing the appearance of relief.

For painting raised areas, you can use any combination of brushes suited to the size of the details. Decorators use hog-hair fitches but artists' brushes are just as good. You can do one of two things when picking out with paint. The first is to choose a tone slightly darker than the main color you choose for the raised area. Brush this darker color into recesses or background sections of the raised detail and, when that is dry, go over the higher relief with the main, lighter color, so that the darker shade remains only in the crevices. This approach is easier if the lighter

Left page The positive method of picking out. Gilded cornices may appear a little grandiose for most domestic interiors but if handled carefully, with the gilding dulled, they can give a sense of warmth and richness from their reflective surfaces.

Above and left The negative or reductive technique of picking out. Brushing on glaze and then wiping it off raised surfaces so that it remains in crevices is not only easier, it is a more subtle way of highlighting areas to enhance them. The effect is also more delicate.

131

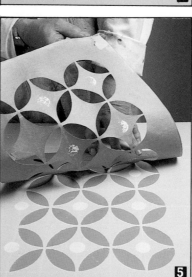

color is opaque. Alternatively, if the paint is thinner and more translucent, you can paint overall in the lighter color and then outline the chosen areas with the darker. The first method is rather quicker than the second.

Method II Here, glaze is used to emphasize the same principle of light and shade, but in a considerably subtler manner than with paint. Glaze is also easier to apply. If the whole of the molding area is finished in an oil-based paint, it can be left to dry and then coated with an oil-based glaze, tinted with oil paint to a darker tone. The glaze can be stippled thoroughly into the recesses. Then, with a rag wrung out in mineral spirits, the glaze can be wiped straight off the raised areas, leaving the darker color in the crevices to emphasize the shadowy areas, thus throwing the details into high relief.

STENCILING

Stenciling, like marbling and wood graining, is a technique that dates from antiquity. There are examples of it on surfaces from walls to shields, domestic houses to painted tombs. Its simplicity is equalled by its charm, and, on occasion, by its beauty, because it can be applied with great delicacy as well as with the more familiar bold effect. It can also be executed on almost any painted surface — provided that it is clean and sound — and many unpainted ones, including wood and metal. Its popularity has waned in Europe until recent years, due largely to the use of wallpapers and, in part, to its rather vulgar and heavy-handed use in the late Victorian period in England where it was often misapplied in a banal, grandiose manner without the delicacy and fineness of earlier periods.

In Europe, stenciling has tended to be more of an industrial process for much of this century, appearing on signs, aircraft and roads rather than in domestic interiors; sculptors have used it to the full on painted structures for its strength and crispness, but interior designers have only recently warmed to it, largely inspired by the work of one particular individual, Lyn Le Grice.

In America, it was better used from the start, and indeed the technique was largely preserved by American settlers who turned it to unpretentious good use as a form of decoration on plaster, floors and furniture in clapboard houses. As a result, it achieved status and has become part of American folk-art and culture.

On pale walls, either in a frieze pattern or as a pattern distributed from the corners spreading outward into the room, there is no real color limitation, provided that the colors do not become too various and loud or the pattern confused and disjointed. Very muted color can be particularly beautiful — on off-white walls, gray or blue-gray and beige patterning can make the wall look like a big, damasked cloth; or with a simple pattern of strong regularity you can use color of almost Etruscan intensity — such as vibrant Indian red or deep, vivid blue. On woodwork, if it is clear and sealed, patterning of muted, warm earth colors and olives looks very well, giving that variety within unity that one sees in high quality marquetry. Woodwork that is painted overall can take areas of very intense color; provided that it's not overdone, it can be extremely effective.

As with most other decorative effects, it is always better to err on the side of restraint in stenciling; you can always strengthen what you've done, but it's very hard work toning it all down again, let alone taking out whole sections of a design. Strong patterning works well on floors, where you can "paint your own rug" if you wish; you can, in fact, use stenciling anywhere in any room — provided that you protect it with varnish. With this technique, you also have complete control over the design. You can choose one from any source; you will simply need to scale it up by tracing it, then drawing a grid over the tracing and scaling up the squares. It is possible to buy ready-cut stencils with modern or traditional designs, if you wish. It is then easy and inexpensive to test your designs and plans, either by using colored paper cut to the pattern and stuck to a wall or other surface with masking tape, or by painting the pattern onto a piece of lining paper and pinning that to the surface to gauge the effect. You need apply not one drop of paint until you've got the whole thing absolutely right, which is a great luxury. Stenciling is a process where the preparation takes longer than the application. However, this is no drawback as half the creative enjoyment of the process lies in the design and preparation of the stencils themselves.

7

Stenciling sequence

1 Fix the stencil-board firmly to the surface to be painted.
2 Apply paint vertically with a flat-ended stencil brush, using a pouncing action.
3 Press the stencil-board flat if necessary to stop paint getting under the edges.
4 When the first stage of the pattern is dry, apply the next.
5 Lift the stencils vertically, peeling backward, never sideways.
6 A basic shape carefully aligned on a grid can alternate various orders of pattern.
7 This pattern can be touched up using an artists' brush or sign writers' "pencil".
8 This basic stencil can be used in a series of variations.

8

■ **Cutting stencils** Stencils can be made from a choice of two materials. You can either use clear acetate or oiled stencil-board. Both have their advantages and both require slightly different methods of preparation. If you use acetate for stencils, you will need a special pen to draw on it, a technical pen; for oiled board, a felt-tip pen. A scalpel is necessary in both cases, and a cutting surface — hardboard, plywood, chipboard or, best of all, glass. If you are going to copy a design, you will need tracing paper, and with oiled board you will also need carbon paper and a 5H pencil or fine-gauge knitting needle. In both cases, you need a straight-edge, preferably a metal ruler, and masking tape.

Acetate Using clear acetate for stencils bypasses the need for tracing paper. You can place the acetate directly over the design and then trace the design straight onto the acetate with a technical pen. This is particularly useful if you are going to work in a variety of colors and need a separate stencil for each color area. Acetate is most effectively cut on glass, because previous score marks on a cutting board can jolt or turn the blade, and acetate can split if cut suddenly at an awkward angle as a result of snagging the blade like this. Fix the acetate in position on the glass with masking tape to stop it sliding, and protect the edges of the glass with tape to prevent it chipping or cutting you. Hold the knife firmly and cut smoothly and steadily. Cut toward yourself, but *never* place your hand in the path of the advancing knife blade; a scalpel, even moving slowly, will almost always make a deep wound. When cutting curves, turn the board steadily, not your knife hand. Cut small, detailed areas first, but don't try punching them out on glass. Any rough edges can be smoothed with fine abrasive paper. The drawback of acetate is that it has a habit of curling up as you cut it, owing to the heat of your hand, and it can crack and split on tight curves. In general, though, it is rather easier to cut than oiled board.

Oiled stencil-board This has the advantage of being a lot cheaper than acetate, and thicker; it is possible to bevel the edges of the board as you cut it to ensure that paint doesn't seep beneath the rim when you apply the stencil. When cutting the stencil from board, copy the chosen design with a fine felt-tip pen onto tracing paper, then transfer it onto the board, using the point of a fine-gauge knitting needle or hard 3H or 5H pencil and carbon paper. Leave a margin of at least 1 – 2½ in (2.5 – 6.2cm) around the design, to ensure that the stencil isn't floppy. As with acetate, fix the board to the cutting surface with tape and cut it in the same manner (preferably on glass). Lining up board stencils is a little trickier than it is with acetate. The best way is to cut them all first and then align them exactly one on top of the other, trimming all the boards to the same size. Also, make small holes in the corners so that you can put a small pencil mark through them onto the wall in order to guide the position of the stencils in the sequence.

Make sure when you cut a motif

Above Subtle and graceful stenciling, applicable to many interiors, adds quiet panache to an upright piano.

Above right An elaborate stencil on a floor surface, abuting marble. A somewhat eccentric mixture but the coloring is well balanced.

Lower right A stenciled chest of strong, bold design is excellently in keeping with the chunky, four-square stencil of the stripped and varnished floor, which has the eye-catching simplicity and panache of good industrial code designs.

that you always cut it from one piece of board; don't continue it into another piece as the join will always show when you paint in the pattern. Never cut too near the edge of the stencil boards or they won't be rigid enough to use and may even break up. Also, make sure that if your design goes around a corner, you cut two stencils that allow for this; don't bend a stencil around the corner — it may work once but the second time you may get seepage of paint, or worse.

■ **Paint and brushes** You can stencil on practically any type of paintwork, provided that it is clean and in good condition. Gloss paint has less key than the other types and so is marginally less suitable, but you can put a matt varnish over it if you don't care about losing the shine. Natural wood needs two thinned coats of matt or satin varnish to seal it, but any other paint surface will take stenciling, provided that it is level.

Similarly, there is a wide choice of stenciling paint. If you are in a hurry there are sign-writers' colors, thinned with matt varnish or mineral spirits; artists' acrylic colors, thinned if necessary with acrylic medium or water; and latex paint, tinted with artists' acrylics or universal stainers. These are all very quick-drying. For woodwork, flat oil-based paint, undercoat or egg-shell, tinted with artists' oils and thinned with mineral spirits, are all very suitable but rather slow-drying, though they can be speeded up with a drying agent. In all cases, never make the paint thinner than half and half because if it is watery it will run under the stencil and mar the pattern.

Somewhat surprisingly, perhaps, sprays are not very good for stenciling. If you do use sprays, the template must be held down very firmly around the area, for sprayed paint has a marked tendency to seep. If you obey the manufacturers' instructions and deliver the spray horizontally, not at an angle, you lessen the risk of seepage but it can still happen. Sprays are also far more expensive than other mediums, and over an area as small as a stencil it's difficult to get solid, even coverage without sweeping back and forth over the area; this means that you get either seepage or, if you hold the spray close to the surface for denser coverage, drizzling. If you hold the spray well back and don't go heavily into the edges of the motif, then you get a fuzzy, fluffy shape instead of sharp definition. In some

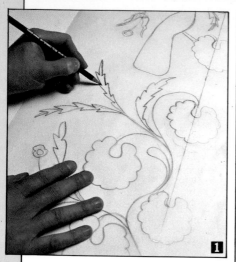

Preparing a stencil

1   Trace the design onto the stencil-board. If you don't wish to use carbon paper, pencil the design on both sides of the tracing paper and go over it to transfer the graphite on the back of the paper onto the stencil-board.
2   These faint lines can be gone over afresh with a sharp pencil.
3   You can trace a design directly onto clear acetate and cut it immediately; on the other hand, acetate cracks easily when you cut it and it is floppy when held vertically against a wall; this allows paint to creep under the edges.
4   When painting, pounce the flat-ended brush vertically up and down, and try not to overload it.

circumstances that might be acceptable but not usually; it gives the appearance of a faded decal. Also, you don't save time because you have to go back and forth gradually over the same region, which takes a good deal longer than one good application with a stenciling brush.

Stenciling brushes closely resemble shaving brushes with the bristles chopped off short and square. They are stubby and squat and sometimes known as 'pounce' brushes after their method of

application — an up-and-down pouncing or dabbing stroke. They aren't very expensive and there's no need to make a substitute — cutting down another brush isn't necessary and won't be any cheaper. They come in various sizes, with the smallest ones resembling fitches; some are actually called fitches. Fitches are round in section, flat-ended and used for very detailed stenciling. All of these brushes leave an 'orange-peel' texture on the surface, a slightly grainy appearance. If you really don't

want this texture, use a sponge, an ordinary decorating brush, a folded rag or paint pad. An ordinary brush carries the risk of coating slightly under the edge of the motif. Whatever tool you use, the secret is not to overload it. Test on a piece of paper with a test stencil to gauge the correct loading.

Other tools You will also need a hardboard or, preferably, glass panel as a palette for the paint, as it is not advisable to put paint straight on the brush from a container; clean rags, appropriate solvent, masking

tape, spirit-level, a straight-edge, chalk, T-square and plumb-line.

■ **Application** First, you must mark the position of the stencils on the wall. If your design follows a frieze format — that is, a series of patterns following each other along a regular plane, usually at eye level or above — never take your measurements for verticals and horizontals from the walls or ceiling; they are usually out of true. Draw all horizontal lines in chalk using a spirit-level, and take verticals from this horizontal using a T-square. Then check these with a plumb-line. Next, mark the position of each stencil on the wall with a pencil, using corresponding notches on the stencil to line it up; that is, the pencil mark on the wall should fit into the notch on the stencil.

If you have a freer overall pattern on a floor, or a rug style, then square off the floor with chalk lines and mark the position of each stencil as before. If you have a border around the pattern, it should be

Above left *Cut a different stencil for each color in complicated designs.*

Left *To imply age, soften the color by adding a transparent glaze mixed with a little white pigment to the color.*

Above *Mellowed coloring improves many simple, naive folk designs. A coat of tinted varnish grays the greens, makes reds pink and turns yellows a gentle amber.*

When you come to apply the paint, fix the stencil firmly in position with masking tape. Pour out a small amount of paint onto the palette, dip the face of the brush once into it, and then stamp it firmly out on a clean area of the palette or, preferably, another testing surface, to distribute the paint evenly and avoid any heavy or unequal loading. Then, working from the edge of the motif inward to the center, dab the thinly coated brush straight onto the surface. Rock it back and forth slightly to transfer color evenly, but don't smear it across the surface or you'll risk carrying paint under the stencil. When the motif has been filled in, let the paint set for about 30 seconds and then lift the stencil off carefully and *vertically* and transfer it to the next position. If the board is going to overlap adjacent stencils, apply them alternately to avoid smudging the paint.

If you are working in several colors, always allow one color to dry before commencing the next. Don't worry about any build-up of paint on the edges of the stencil as this just means you are applying the paint correctly; but clean it off at intervals by wiping it with a sponge or rag steeped in the appropriate solvent, in case it begins to get badly clogged. When you've finished, clean the stencils and brushes with the same solvent and store the stencils flat. Separate them with tissue paper or baking foil; it's a good idea to keep used stencils because if ever you wish to repeat the pattern it will save you a good deal of time and work. It also means that you can retouch the surfaces at any time, if necessary; stencil work can be retouched with relative ease. Stored properly, stencils last a lifetime.

Remove any pencil and chalk marks from walls with a clean eraser. This is important as you will have to varnish stenciling; any graphite dots will show up under varnish like little hazy flies because varnish tends to disperse light through its surface and so exaggerate blemishes that would otherwise be unnoticeable. Leave the paint to dry for at least a day — two days for oil-based paints — before varnishing. Two coats of matt or satin-finish varnish are a good idea on walls, and on woodwork and floors they are absolutely essential. Floors look best with satin or matt over gloss varnish to give them a soft sheen, and need at least three coats, preferably five.

Above *The simple stencils on these kitchen units achieve a gentle, homey effect due to their muted coloring and regularity. The use of lining adds greatly to their positioning; without it they'd seem to float in space. Lack of lining would demand more patterning; as it is, there's enough.*

Above right: Glass painting
In the nineteenth century, painting glass was frequently used as a substitute for the slower, more costly process of true stained glass. When glass painting followed the basic forms of stained glass design, it was often most successful. Much 19th-century "stained glass" is actually painted. When the paintwork became looser in form, the result was often weak and disordered. In this example, the strength and steadiness of the flowing, painted flower-and-stem tracery form a graceful harmony with the genuine stained glass panels that surround the painted design.

equidistant from the wall all round. Lay out the border by drawing parallel chalk tracks; the distance between the lines should be the width of the stencil templates. Of course, it could be that your room really is very off-square; the best thing to do in that case is to draw a right angle where your border chalk lines meet at the corner, and then when you see just how crooked the room really is, draw the border so that it is somewhere between the two. For example, if the corner of the room is about 41°, and the true corner angle 45°, draw your border at about 43°. Also remember that you can't take motifs round a corner on a flat floor in the same manner as you would on the vertical corner of two walls. You must divide the length of your border by the number of motifs you have and space them out evenly so you don't get one motif colliding with another on the corner turn. Mark the central point of each motif on the stencil template and the tracks.

GLASS PAINTING

Traditionally, the Japanese have been the finest exponents of painting on glass, the lightness and translucency of their brushwork being admirably suited to the technique of placing a thin, colored liquid on a very thick, rigid and transparent one. In Europe, the Victorians showed a considerable interest in this skill, but only a slender facility for using it. Their approach was an unhappy marriage of minute execution of Dürer-like intensity to the stylized rigidity of stained glass, and the result was rather like two-dimensional taxidermy. This unfortunate result reveals that glass as a painting base offers two distinct options: it can either be used in the manner of medieval stained glass, which makes the most of the light-filtering quality of the medium and makes no attempt to emulate easel painting; or, on the other hand, it can be used for the delicate miniaturism and subdued coloring that has more to do with glass engraving. Painting on glass, as far as interior decoration is concerned, is therefore better orientated to either one or other of these approaches — not both together. However, both methods have a certain amount in common with stenciling, which is relatively straightforward on glass.

To produce a stained-glass pattern, a design can be worked out in the same manner as it is for stenciling, whether or not it is copied from another source. The design should be drawn out on paper or card, the glass simply placed over it and the design traced either in paint or wax pencil straight onto the surface of the glass. The best approach is to trace in the outline with an artists' sable brush, using the appropriate color for each section, and then filling in the area with the right colors.

There are paints expressly designed to be used on glass, with an opaque finish that lets light through evenly, but their consistency varies and one should note the manufacturers' instructions when using them. Artists' oil color works well too, provided that the glass is cleaned thoroughly first, usually with ordinary window detergent, and allowed to dry thoroughly. Degrees of transparency can be achieved with oil, depending on the amount of solvent added, but all glass with a

light source behind it allows illumination of the paint.

The delicate, engraved-glass appearance of finer images can be traced off in the same way, or taken straight off an enlarged (or reduced) photograph. A result of extraordinary delicacy is possible by laying a monochrome print under glass and working in artists' oil color in either monochrome grays or sepia over the image, filling areas on the glass to correspond with those on the photograph. The result — when light shines through it — is of a monochrome engraving. It is best to pursue this technique in these colors; to use a large variety of colors can look rather unpleasant — somewhat kitsch — as though you'd stuck a large number of colored decals on the surface, which have started chipping and peeling. But in subdued grays and sepia this never occurs. It is easy to remove unwanted smudges with a cloth or sponge soaked in mineral spirits. If the piece has already dried, paint can be removed, if necessary, with a palette knife; keep the knife as flush to the surface as possible. Glass painting is best protected by placing it under another piece of glass.

Stenciling a Toy Chest

The first decision to make is whether you want to achieve the type of images on page 119, or those on page 133. If you choose the latter, remember that putting a stencil over a curved lid surface is awkward if the stencil is big, so use floppy acetate if possible or build the images up from small stencils, if using card. A vivid, narrative design on flatter areas needs color control. Decide on two base colors for sky and ground, and on where their dividing line will fall, to hold the design together. Children respond to bright, particularly primary colors and to characters they recognize. Images are often repeated but in different situations. Stenciled figures are always best reduced to simple color sections; most cartoon figures lend themselves to this, as animation has similar requirements. The different color sections can also be swapped over for different figures. You should always protect the surface with varnish.

GLOSSARY

alive
A paint or varnish surface that is still soft, wet and in a workable state. When it begins to set and becomes tacky, it ceases to be alive .

base coat
The paint on top of which other layers are applied, i.e. undercoat.

base color
The foundation or background color of a design, on occasion also called the ground color. Off-white, for instance, is the base color of white Sicilian marble.

cissing
An effect created by spattering solvent on a paint or varnish surface before it is dry. Mineral spirits are usually spattered on oil-based paint, and water on latex, but the method works well the other way round, too. round, too.

claircolle
Untinted distemper. Size and whiting mixed and applied without color to give a simple, even, off-white base to colored distemper.

crossing off
The final finishing stroke of a *paint* method. The term is interchangeable with "laying off", except that the latter tends to refer to the whole activity and the direction of the crossing off, whereas crossing off refers to the single action. Crossing off *varnish* means making a stroke at right angles to the rest of the varnish strokes — which are usually vertical — at the bottom of a panel or surface. This stops the varnish running.

cross-stroke
A criss-crossing stroke in an "X" formation, usually employed when laying on gloss paint, distressing color wash, or for other broken color techniques.

cutting in
Painting into an angle or onto a narrow surface like a glazing bar. A cutter is a brush of moderate or narrow width with a flat, chisel shape or hatchet point (*see chapter two*).

diffuser
A T-shaped pair of tubes used by artists to diffuse paint or ink in a very fine spray. The base of the "T" is placed in the liquid; suction is created in the tube by blowing in one arm of the tube, drawing the paint upward, and a second breath disperses the liquid spray from the other arm.

flatting-oil
A 1:6 mixture of boiled linseed oil and mineral spirits, used to "float" color on (*see "Marbling"*).

gesso
A hard, off-white surface traditionally made from hide-glue (usually rabbit skin) and whiting, and often used as a base for gilding work. The best is still made this way, and is applied hot but never boiling in layers, mostly to picture frames and furniture. A cold, ready-to-use liquid form is also available and is reasonably adequate for most decorative work.

goldsize
A size used in the gilding process. Its clear, viscous quality thickens paint.

ground
Any surface to which a paint finish is to be applied. The term is also sometimes used to mean the base color; for example, white Sicilian marble has an off-white ground.

gypsum-board
A form of plaster board (*see below*).

jamb duster
A large, pliable, flat decorators' brush which can be used as a substitute tool for the dragging and combing technique.

laying off
Another term for crossing off (*see above*). With paint, it means the uniform direction of finishing strokes as well as the application of them; with most types of paint on most surfaces, it should be done toward the light.

laying on
The action of applying paint or varnish initially, before any finishing strokes. Laying on may be in any direction unless specified, but should never be treated as a brush method for achieving a finished surface, except when you are distressing a color wash (*see "Broken Color"*).

masking tape
A strong, sometimes plastic-coated, sticky-backed tape used for blocking off surfaces to protect them from paint, and for holding other materials, such as paper and polythene, in place.

molding
Raised relief patterns in wood and plaster. The parallel raised surfaces of doors and architraves, windows and ceiling surrounds are also loosely referred to as moldings.

mottle
An irregular pattern similar to that cast by the shadows of leaves. The mottles can be of any size from fist to pin-head. A mottling action is a smudgy stipple created with a brush or sponge by a dabbing action up and down, or with a spray by making irregular patches with random jets.

over-brushing
Brushing one color loosely on top of another, either to cover it partly or to alter its tone.

over-painting
Covering and obliterating one color with another.

oxide of chromium
A deep gray-green color, between moss and mid-olive, similar to terra verde. Originally made from chrome oxides.

pattern-work
The general structure and flow of a design, as in stenciling, on moldings and lining, or decorative work of any kind.

plaster board
A sheet of plaster with a stiff paper covering on either side. One side of the board is finished and can be used without further treatment.

plumb-line
Traditionally, a cord with a lead weight at the bottom. The plumb-line is held steady like a stationary pendulum to test a true vertical.

pumice
A light, porous, volcanic rock used for polishing and scrubbing. Now usually available in a lump approximately 5in (13cm) across, it should be cut in half (with an old saw as it blunts the teeth) and rubbed on a wet flagstone before use. It is excellent for washing down shiny surfaces. Powdered pumice should be mixed with oil and applied with a piece of thick felt to achieve a smooth finish on paint or varnished surfaces. After using it, the surface should be rubbed with a lint-free cloth and washed in turpentine. Pumice gives a soft, deep gleam.

retarder
A colorless gel agent mixed with some water-based paints, like acrylics, to slow their drying time by inhibiting evaporation.

rotten-stone
Decomposed limestone, powdered for use as a polish. Used with a lint-free rag and linseed oil, this lubricant should be rubbed over a surface after a felt-rub with pumice, sanding or an application of waterproof abrasive paper. The surface should then be polished with dry flour. The result is a soft gleam.

spirit-level
A wooden or steel rod with a phial in the center filled with spirit, used for testing a true horizontal. When the bubble in the liquid is in the exact center of the phial, the surface is exactly level.

steel wool
This is available in many grades from coarse to very fine, and looks like a bundle of gray hair. Used dry, or with water or oil, it is highly effective for rubbing down paintwork, and giving a key or tooth to smooth surfaces such as aluminum or galvanized steel. It is suitable for removing rust, used in conjunction with mineral spirits. *Don't* use it on surfaces where you intend to apply water-based paint, as embedded particles of it may cause rust under latex. In this case, use bronze wool.

stippling
Using a sharp, stiff-bristled brush to create a pattern that resembles fine grit, either by exposing a color used under glaze, or by applying one color on top of another.

straight-edge
A long ruler. Any stiff, straight object which offers an undulation-free edge as a guide.

tack-rag
A sticky cloth designed to pick up greasy, gritty particles from a surface (*see chapter three, "Newly Stripped Wood"*).

template
The master- or guide-drawing for a design, from which stencils are cut and tracings made.

texture (tactile and visual)
Literally, the way a surface feels, usually to the touch. However, with paint, the visual texture is often more important. This means the impression made on the eye; a surface can evoke roughness without actually being rough. The term "surface texture" refers to whether it is actually rough or smooth to the touch.

tipping off
Touching the tips of brush bristles on the side of a paint kettle to eject excess paint from them.

translucent color
Color which allows you to see the presence of another beneath, in the manner that gauze allows you to see through it.

transparent color
Color which alters the tint of another beneath it without obscuring it, as tinted spectacles alter the color of the sky without obscuring it.

T-square
A T-shaped ruler. The right-angled cross-bar is placed against the straight side of a surface, ensuring that a line drawn along the remaining long rule will be at right-angles to it.

under-toning
The tones of background colors. A soft ocher mottle applied to marble before the veins and top glazes are applied is under-toning.

wash
A thinned coat of color applied all over a surface, either over another paint or as a mist coat when latex is applied to new plaster.

water tension
The ripples caused in water-based paint where patches of it dry at the edges and don't blend into the adjacent patch evenly. This is caused by evaporation from the thin edge of the paint area being faster than that from its center.

wet-and-dry paper
An abrasive paper designed to smooth a surface with alternate applications of lubricated and dry abrasion.

wet-edge
The alive (*see above*) edge of a paint area. The wet paint into which the paint of an adjacent area can be harmoniously brushed.

INDEX

Page numbers in *italic* refer to captions and illustrations

A

acrylics, bamboo effects, 117
    tinting with, 23
aerosols, 25
ageing techniques, 80
alkaline strippers, 36
aluminium primers, 42, 43
antiquing, 80, *81*
artists' acrylics, bamboo effects,
    tinting with, 23
artists' gouache, tinting with, 23
artists' oils, lining, 121
    tinting with, 23
artists' powder pigments, tinting
    with, 23

B

bamboo finish, 116-17, *116-17*
beer graining, 108-11
bird's-eye maple, *109*
black serpentine marble, 93
blow-lamps, *25*, 37-8, *38*
blow-torches, *25*, 38, *38*
blue marble, 97
broken colour, 51-80
    antiquing, 80, *81*
    colour washing, 53-6, 80
    combing, 62-70, *66-9*
    dragging, 62-70, *63-5*, *68-9*
    rag-rolling, 70-8, *70-7*
    ragging, 70-8, *70-5*
    shading, 56-8, *56*
    spattering, 78-80, *78-9*
    sponging, *57*, 58-60, *58-60*
    stippling, 61-2, *62*
    thinning densities, 53
brushes, *22-3*, 24-5
    cleaning, *20*
    loose hairs, *20*, 28
    restoring, 30
    stencilling, 136
    types, *22-3*
burning off paint, 28, 37-9, *38*

C

canvas, preparing, 47
ceilings, painting, *26*, 27
    shading, 44, 56
chemical strippers, 36-7, *37*
chrome, painting, 48
cissing, 86
clothing, removing paint from, 30
cloudy effects, 61
cold colours, 14
colour, cold, 14
    discordant, 16
    hues, 10
    mixing, 11-13, 16
    off-hues, 15-16

room's aspect and, 14-15
    shades, 10, 11
    theory, 10-13
    tints, 10, 11
    tonal values, 10
    tones, 10-11
    transparency, 12
    use of, 13-16
    warm, 14
    wavelengths, 14
colour schemes, choosing, 9-17
colour washing, 53-6, 80
colour wheels, 10, *10*, 11
combing, 62-70, *66-9*
complementary colours, 10-11, *11*
    mixing, *10*
copper, painting, 48

D

discordant colours, 16
distemper, application, 54-5
    colour washing, 53-4
    making, 54
    painting over, 35
    removing, *34*, 35, 48
distressed effects, *54*, 55-6
doors, painting sequence, *21*, 27
    tortoise-shell, 102
dragging, *12*, 62-70, *63-5*, *68-9*

E

egg-shell paint, 20-1
    thinning, 53
emulsion paint, 20
    broken colour-work, 53
    dragging, 66
    on floors, 68
    laying off, 28
    shading, 58
    stripping, 36
    thinning, 53
equipment, 24-7
    brushes, *22-3*, 24-5
    rollers, *24*, 25
    spray guns, 26-7
    for stripping, *25*
    unorthodox, 27

F

fabrics, preparing, 47
fading, 31
fantasy decoration, 83-117
    marbling, *84-101*, 85-100
    porphyry, *105*, 106-7
    tortoise-shell, 101-6, *102-4*
    wood graining, *106-7*, 107-16,
        108-15
ferning, 65
fillers, for plaster, 39, *40*

finishing touches, 119-39
    glass painting, *138*, 139
    lining, *120-3*, 121-4
    picking out, *130-1*, 131-2
    stencilling, 132-8, *132-8*, 139
    trompe l'oeil, 124-31, *124-9*
flat-oil paint, 20
    colour washing, 55-6
    on floors, 68
    shading, 56-8
    thinning, 53
floors, 98-101, *100*
    combing, 68-9
    marbling, 98-101, *100*
    paints for, 100-1
fog coat, 41
furniture, antiquing, 80
    cissing, 86
    dragging on, *65*
    lining, 124
    metal, 47
    ragging, 74
    sponging, 61, 62
    spraying, 25
    stencilling, *134*, *135*, *138*, 139

G

gilding, *131*
glass, painting, *138*, 139
    preparing, 43
glazes, *15*
    graining, 111-16
    painting, 12-13
    picking out, 132
    thinning, 53
gloss paint, 20, 21
    drying, 28
    flaking, 29
    laying off, 29
    removing, 40
gouache, tinting with, 23
graining, *106-7*, 107-16, *108-15*
green marble, 94-7, *94-5*

H

hot-air blowers, 38-9
hues, 10

K

kettles, paint, *21*, 25
knotting, 42-3

L

lacquer, removal, 44
    removing, 40
laying off, 28
light, mixing, 13
    prevailing, 10

quality of, 14
wavelengths, 14
lighting, 30, 31
lining, *120-3*, 121-4
   furniture, 124

M

mahogany, *115*
marbling, *84-101*, 85-100
   black serpentine marble, 93
   blue marble, 97
   floating, 98
   floors, 98-101, *100*
   green marble, 94-7, *94-5*
   paper rocking, 98
   red marble, 93-4
   rose marble, 90-3, *90*
   terra verde marble, 94-7, *94-5*
   varnishing, 94, *98*
   white Sicilian marble, 89-90, *98*
   yellow marble, *96-7*
masking, 29
metal, preparing, 47-8
mid-sheen paint, 20-1
mist coat, 41
mouldings, lining, *123*

O

oak graining, 111
off-hues, 15-16
oil glazes, antiquing, 80
   dragging and combing, 65-6
   ragging, 76
   sponging with, 59
oil-based paint, dragging and
      combing, 65-6
   ragging, 76
   shading with, 57-8
   sponging with, 59-61
   types of, 20-1

P

paint, burning off, 28
   fading, 31
   floor, 100-1
   mixing, 16
   painting over, 35
   quantity needed, 28
   stencilling, 135
   stripping, *25*, 28, 35-9
   tinting, 21-4
   types, 20-1
paint kettles, *21*, 25
painting, preparation, 33-48
   sequence of, *21*, 27
paper, stripping, 44-7
paper rocking, 98
peeling, 29
picking out, *130-1*, 131-2
pigments, tinting, 21-4
plaster, defective, 39, *40*
   filling, *48-9*
   new, 20, 39-41, *41*
porphyry, *105*, 106-7
   cinnamon, 106

grey, *105*
   pink, 106
poster colours, tinting with, 23
powder pigments, tinting with, 23
preparation, 33-48
primary colours, 10, 11
primers, for wood, 42, 43

R

radiators, 30
   preparing, 48
   spattering, *78*
rag-rolling, 70-8, *70-7*
ragging, *12*, *16*, 70-8, *70-5*
red marble, marbling, 93-4
rollers, *24*, 25
rose marble, 90-3, *90*
rosewood, 111
rust, 48

S

safety, paint stripping, 39
secondary colours, 11
serpentine marble, 93
shades, 10, 11
shading, 56-8, *56*
   ceilings, 44, 56
shellac, removal, 44
Sicilian marble, 89-90, *98*
single-colour rooms, *13*
spattering, 78-80, *78-9*
spirit strippers, 36
sponging, *57*, 58-60, *58-60*
   furniture, 61, 63
spray guns, 26-7
spraying, 29, 30
   furniture, 25
   masking, 29
stained glass effects, *138*, 139
stainers, tinting with, 24
steamers, paper stripping, 45-6, *46*
stencilling, *14*, *112*, 132-8, *132-8*, 139
   application, 137-8
   brushes, 136
   cutting stencils, 134-5, *136*
   furniture, *134*, *135*, *138*, 139
stippling, 61-2, *62*
stone, stripping paint from, 37
stripping paint, *25*, 28, 35-9

T

tack rags, 43
terra verde marble, 94-7, *94-5*
tertiary colour, 11
thinning densities, 53
tiles, preparing for painting, 48
tinting, 21-4
tints, 10, 11
tonal values, 10
tones, 10-11
tortoise-shell, 101-6, *102-4*
   amber, 104
   auburn, 104, *104*
   finishing, 104-6
   golden, 102-4, *102-4*

trade egg-shell, 21
transparency, 12
trompe l'oeil, 124-31, *124-9*

U

undercoat, 20
   thinning, 53

V

varnish, 31
   stripping, 36, 43-4, *44*
varnishing, 24
   marbling, 94, *98*
   over colour washes, *55*
vinegar graining, 108-11
vinyl papers, stripping, 46

W

wallpapers, stripping, 44-7
walls, antiquing, 80
   colour washing, 80
   dragging on, 66-7, *69*
   new plaster, 39-41, *41*
   old plaster, 39, *40*
   painting, *26*, 27
   preparing, 39-40
   rag-rolling, 73
   tortoise-shell, 102
walnut, 111
warm colours, 14
washable papers, stripping, 46-7
water-based paint, colour washing
      with, 53-5
   shading with, 58
   types of, 20
wavelengths, light, 14
waxed wood, preparing, 43-4, *44*
white Sicilian marble, 89-90, *98*
windows, 30
   metal frames, 48
   painting sequence, *21*, 27
wood, new, 41-3, *42*
   preparing for painting, 41-3, *42*
   varnished, 43-4, *44*
   waxed, 43-4, *44*
wood graining, *16*, 106-7, 107-16,
   108-15
   beer graining, 108-11
   bird's-eye maple, *109*
   glaze graining, 111-16
   mahogany, *115*
   oak, 111
   rosewood, 111
   vinegar graining, 108-11
   walnut, 111
woodwork, antiquing, 80
   dragging, 67-8
   integration with walls, 14
   painting sequence, 27
   stripping, 36

Y

yellow marble, *96-7*

ACKNOWLEDGMENTS

The photographs on these pages were reproduced by kind courtesy of the following:

6 Elizabeth Whiting & Associates; **9**(inset) Karen Bussolini; **11**(top) Dulux; **11**(bottom) Smallbone of Devizes; **12-17** Elizabeth Whiting & Associates; **33, 49** Home Improvements Guide; **51**(inset), **56-7, 65, 71** (top), **75, 78**(top) Elizabeth Whiting & Associates; **81** Martex, West Point Pepperell; **83**(inset), **85, 91, 100, 108, 110, 112** Elizabeth Whiting & Associates; **117** Royal Pavilion, Brighton; **119**(inset), **120, 124-9, 134, 135**(bottom) Elizabeth Whiting & Associates; **135**(top) Karen Bussolini; **138** Smallbone of Devizes; **139** Elizabeth Whiting & Associates.

All other photographs are the property of Quill Publishing Limited.

While every effort has been made to acknowledge all copyright holders, we apologize if any omissions have been made.